[英国]戴维·C.卡特林 著 郝记华 张晟霖 译

牛津通识读本·

天体生物学
Astrobiology
A Very Short Introduction

译林出版社

图书在版编目（CIP）数据

天体生物学 /（英）戴维·C.卡特林（David C. Catling）著；郝记华，张晟霖译. -- 南京：译林出版社，2025. 3. -- （牛津通识读本）. -- ISBN 978-7-5753-0506-8

I. Q693

中国国家版本馆CIP数据核字第2025RX0890号

Astrobiology: A Very Short Introduction, First Edition by David C. Catling
Copyright © David C. Catling 2013
Astrobiology: A Very Short Introduction, First Edition was originally published in English in 2013. This licensed edition is published by arrangement with Oxford University Press. Yilin Press, Ltd is solely responsible for this bilingual edition from the original work and Oxford University Press shall have no liability for any errors, omissions or inaccuracies or ambiguities in such bilingual edition or for any losses caused by reliance thereon.
Chinese and English edition copyright © 2025 by Yilin Press, Ltd
All rights reserved.

著作权合同登记号　图字：10-2018-429号

天体生物学　［英国］戴维·C.卡特林／著　郝记华　张晟霖／译

责任编辑	许　昆
装帧设计	景秋萍
校　　对	王　敏
责任印制	董　虎

原文出版	Oxford University Press, 2013
出版发行	译林出版社
地　　址	南京市湖南路1号A楼
邮　　箱	yilin@yilin.com
网　　址	www.yilin.com
市场热线	025-86633278
排　　版	南京展望文化发展有限公司
印　　刷	江苏扬中印刷有限公司
开　　本	890毫米×1260毫米 1/32
印　　张	9.5
插　　页	4
版　　次	2025年3月第1版
印　　次	2025年3月第1次印刷
书　　号	ISBN 978-7-5753-0506-8
定　　价	39.00元

版权所有·侵权必究

译林版图书若有印装错误可向出版社调换。质量热线：025-83658316

序 言

潘永信

在数千年悠久的人类文明长河中，祖先们一直没有停止过思考我们是谁、从哪里来、是否孤独等哲学问题。这些哲学思考在民间形成女娲造人等多彩的神话故事，丰富了先人的精神世界。随着近现代科学观察和研究的深入，人们对于上述问题的认识已不再仅仅停留在哲学思辨，还不断产生客观的观察记录和理性归纳。大家熟知的"生物演化论之父"达尔文，也是科学地设想早期地球如何孕育出最早生命的先驱之一；一提到生命起源，您也会自然地想到著名的米勒-尤里放电实验；诸多教科书或者自然博物馆往往会展示令人着迷的、由生物化石记录的演化历史；中学课本中提到列文虎克发明了显微镜，第一次真切地观察到多种多样微米尺寸微生物的存在；进入太空时代，人们跟随着卡尔·萨根的视角，从遥远的外太空回望了一眼我们共同生活的特殊"暗淡蓝点"——地球。今天，这些科学观察和研究归纳形成了新兴的交叉学科——天体生物学。

在本书中，美国天体生物学家戴维·C.卡特林教授用诙谐幽默又通俗易懂的语言，结合很多亲身经历的趣事，为读者详细介绍了天体生物学研究的内涵，包括：生命是什么、生命和生境起源及演变、太阳系内和系外生命探寻等。作为在天体生物学领域深耕数十年的世界顶尖学者，卡特林教授在本书中呈现给读者一系列前沿的科学观察和深刻的科学思考。译者是中国科学技术大学的郝记华教授，郝教授是我国天体生物学领域的一位优秀青年科学家。因此，本书不仅适合无地球和行星科学背景的不同年龄阶段读者，作为拓宽知识面的科普读物，也适合地球和行星科学方向的学生和学者，作为了解天体生物学研究前沿的重要参考书。

我国已经规划于2030年前后实施"天问三号"和"天问四号"任务，分别对火星和木星及其冰卫星开展探测，并且有可能在国际上率先实现火星样本采样返回。这些深空探测计划的重要目标之一就是认识地外宜居环境和寻找地外生命。我相信读者会从这本书中预先获得一些相关了解。我很欣慰地看到，最近国内多所高校和研究机构已着手培养天体生物学方面的专门人才。这本书的出版无疑会助力本科生和研究生的培养以及科学普及。我也希望更多的青年学子通过阅读这本书认识和了解天体生物学，并一起领略神奇生命在无穷远方和无尽时间中发生的迷人故事！

中文版序言

每个人可能都想过这些大问题：我们在宇宙中是孤独的吗？我们从哪里来？人类将走向何方？在过去的几个世纪里，这些都是哲学家喜欢思考的问题。但随着技术进步，如今这些问题已经进入了自然科学的范畴，因为我们逐渐可以用客观的数据代替哲学的沉思，从而对这些问题做出有效的评估。这个科学领域就是本书所讲的天体生物学，它囊括了上述古老问题，主要研究宇宙中生命的起源与可能的多样性。

总的来说，天体生物学研究包含四大支柱方向。第一个是地球上生命的起源与演化。迄今为止，地球生命仍然是我们认识生命的唯一实例，而且关于生命的诸多谜团仍待解决，包括生命是如何起源的。第二个是搜寻太阳系中的其他生命。比如，我们的邻居火星和金星上是否存在过生命？抑或是，外太阳系中冰卫星的冰下海洋里有生命存在吗？第三个是对太阳系外围绕着其他恒星运行的行星，即系外行星的研究。当前，我们认识到，银河系中大多数恒星都拥有至少一颗围绕它运行的行星。

其中，有不少系外行星的运行轨道处于其恒星的宜居带中，适宜表面存在液态水。最后，天体生物学的第四个支柱方向是对宇宙中其他智慧甚至技术生命的搜寻。在未来的某一天，我们有可能在宇宙中其他地方找到所谓的"技术标志物"，即技术文明显著改变环境的证据吗？

像世界上其他优秀文明一样，中华民族自古以来一直着迷于探索宇宙的本质。例如，东汉博学多才的天文学家张衡（78—139，以发明地动仪而闻名）已经提出了宇宙的无边界性。大约一千年后，宋元之际的思想家邓牧（1247—1306）在其作品中也推断，无穷的宇宙中应该存在众多人类居住的世界。20世纪50年代，人类进入太空时代，发射了许多人造航天器，对太阳系中的行星和卫星进行了持续的探索。与此同时，超大口径的地面望远镜和空间望远镜正在观察类太阳恒星周围地球大小的行星，识别其大气中是否含有潜在生物圈产生的气体。随着探测器和信号处理技术的不断进步，人们重新燃起了对于寻找技术标志物的兴趣。在自然条件下，细胞生命的起源与演化可能可以产生人等由有机物组成的智慧生命。不同于自然演化过程，人类正在创造拥有人工智能和自主能动性的先进机器人，并且相较于它们的创造者，这些机器人可以存在于更严苛的行星环境中，包括一些没有大气的天然卫星。因此，对于潜在地外技术标志物的搜寻可能不仅限于传统的宜居行星。

但是我们依然好奇，在仰望夜空时，我们面对的究竟是一个了无生机的、受控于已知物理与化学规律的宇宙，还是一个拥有众多居住着奇妙地外生命的行星（也许还有卫星）的喧闹大观园，只不过我们现有的探测技术还无法观测到这些生命的存在？也许，宇宙中的某个角落也存在具有意识，甚至已社会化的生物？对于这些问题，天体生物学家们孜孜不倦，希望在本世纪找到答案。正是在这种背景下，我很高兴这本书可以和占世界人口近六分之一的中国朋友见面。亲爱的读者，我希望你们通过阅读此书，可以领略天体生物学这一年轻但蓬勃发展的学科，跟我一起享受思考生命的本质及其宇宙意义等问题的非凡乐趣！

戴维·C.卡特林
美国西雅图
2024年11月

目 录

第一章　什么是天体生物学？　1

第二章　从星尘到行星，生命之所栖　15

第三章　生命和宜居环境的起源　29

第四章　从史莱姆到史莱克　46

第五章　生命：基因组的延续和适应之道　67

第六章　太阳系里的生命　87

第七章　偏僻的世界，遥远的恒星　118

第八章　争议与展望　135

　　　　索　引　141

　　　　英文原文　153

第一章

什么是天体生物学？

名字背后的故事

"**天体生物学**究竟是什么？！"一名美国特勤局特工向无线电对讲机那头喊道。他刚刚在旧金山附近的美国国家航空航天局（NASA）埃姆斯研究中心检查了一位访问学者的身份。来访者对他说自己正在参加NASA组织的第一次天体生物学科学会议。埃姆斯研究中心有一座简易机场，可为"空军一号"提供安全的着陆点。正是在这条飞机跑道上，2000年4月，比尔·克林顿总统带着他的特勤局随行人员飞抵旧金山湾区。

这名特工会提出这样的问题也算正常，毕竟直到20世纪90年代末，学界才对**天体生物学**的含义达成科学共识——应该很少有外行人或者特勤局特工听说过这个术语。彼时，NASA开始推进一个由埃姆斯研究中心领导的天体生物学研究项目，而我正是该研究中心的一名空间科学家。一开始，我的一些同事并不喜欢"恒星的生物学"这一希腊语字面义。有人嘲笑说，生命是不可能存在于地狱般的恒星之中的。相对来说，一个不那么令人扫兴的解释是，"天体生物学"（astrobiology）中的"天体"（astro）涉及的是恒星（包括太阳）**周围**的生命，或者简单来说，太空中的生命。事实上，很多天体生物学家既关心地球上生命的历史，也关心其他天体环境中的生命。天体生物学家们一致认为，我们应该首先对地球上的生命是如何演化的有一个明确的认识和了解，以便进一步思考地外空间生命存在与否的问题。然而，现代科学的一个惊人现状是：迄今为止，它甚至无法完全回答一个孩子可能会问的生物学问题！地球上的生命是如何起源的？我们对此有一些认识，但是其中的诸多细节仍然是未知的。地球与太阳系的哪些特性使得我们的星球适合生命发展？同样地，我们有一些观点，但仍然不了解很多方面。那么又是什么驱动生命向复杂的有机体演化，而不是仅仅保持简单的形式呢？我们仍然不清楚。

为了填补这些人类认知的空白，天体生物学作为**研究地球上生命的起源与演化，以及其他天体环境中可能存在的各种生命形式的科学分支**应运而生。这是我自己更喜欢的一个定义。NASA 则把天体生物学定义为：**对宇宙中生命的起源、演化、分布和未来的研究**。另一个常见的释义是：**对宇宙中生命的研究，或者对宇宙环境中生命的研究**。在这一范畴下，天体生物学家既探讨"地球生命的过去与未来是怎样的"，也探索"地球之外是否也有生命存在"。

20 世纪 90 年代末，天体生物学作为一个学科诞生，与此同时，四项重大科学进展也横空出世。1996 年，科学家在一块火星陨石中发现了颇具争议的古代火星生命迹象。这块重 1.9 千克的石头由于小行星撞击而被炸离火星表面,最终落在地球的南极地区。无论该研究对陨石中存在微体生物化石的解读(详见第六章)是否正确,这项发现都引发了人们对于地外生命的思考。此外,在过去的二十年里,生物学家证实,一些微生物能在比传统认为的更广泛的环境条件下存活,它们甚至能够在极端的温度、酸度、压力或者盐度条件下**茁壮成长**。因此,我们开始考虑在看似极端严峻的地外环境条件下存在微生物的可能性。第三个发现来自1996年NASA的"伽利略号"探测器所拍摄的被冰覆盖的木卫二的表面图像。这些图像揭示了在木卫二冰面上一些冰块曾彼此渐行渐远,说明在冰层下可能存在海洋。第四项是,自20世纪90年代中期以来,天文学家陆续观察到越来越多围绕其他恒星运动的**系外行星**。上述这些发现表明,生命可能定居在这些遥远的系外行星上,抑或潜藏于如我们宇宙后花园般的太阳系内的其他星球上。我们不禁怀疑:在宇宙中生命是不是普遍存在?

思想史上的天体生物学

尽管天体生物学的概念在20世纪90年代才问世,关于我们在宇宙中是否孤独的思考却可以追溯到几千年以前。约公元前600年,西方哲学之父泰勒斯已经开始赞同生命的**多世界论**的观点。后来,希腊原子论学派(认为物质是由不可再分的原子构成的)的拥趸,从留基伯到德谟克里特和伊壁鸠鲁,也支持这种"多元主义"。约公元前400年,德谟克里特的追随者迈特罗多鲁斯曾写道:"在一望无垠的宇宙中只有一个拥有生命的世界,好比在一片辽阔的原野上只有一株小麦——多么反常而不自然啊！"不过,将古代哲学家的多元主义直接等同于现代设想的火星或者系外行星上存在生命可能并不正确。迈特罗多鲁斯全然不知天上的星星只不过是极远的类似于太阳的物体,而认为它们是由地球大气层中的水汽造成的日常现象。原子论学者们想象中的生命世界存在于一个抽象空间中,与现代的平行宇宙观点很类似。与这些相反,柏拉图(公元前427—前347)和亚里士多德(公元前384—前322)认为地球是唯一有生命的天体且居于宇宙中心,这一观点最终还是占了上风并盛行了一千多年。

文艺复兴时期,天文学家们终于证明了地球是围绕太阳运行的。而随着人们逐渐认识到地球也不过是太阳系诸多行星中的一员,很快就出现了对其他行星上存在地外生命的猜测。例如,德国天文学家约翰尼斯·开普勒(1571—1630,提出行星运动三定律)欣然提出其他星球上可能存在生命的观点。随后,在17世纪末,丹麦天文学家克里斯蒂安·惠更斯(1629—1695)在他的著作《被发现的天上的世界》(1968)中想象太阳系**之外**的生命:"那无数恒星周围环绕着的所有行星啊——它们一定也有自己的植物和动物,不仅如此,还会有智慧生物!"彼时,地外生命论一度非常流行,以至于哲学家伊曼努尔·康德(1724—1804)在1755年写过聪慧的木星人和多情的金星人。

与之相对立的是,一些学者囿于宗教认识,仍然坚持认为地球是独一无二的。一个例子就是剑桥大学的威廉·休厄尔(1794—1866),他在《论世界的多样性》一书中反对其他行星上有生命的观点。从某种程度上来说,这是现代**"稀有地球"假说**的先驱。我们将在第八章讨论这一假说。

到了19世纪末,地球外是否有生命的问题已经被视为一个纯粹的科学问题，但科学研究很快走进了死胡同。乔范尼·夏帕雷利（1835—1910）和帕西瓦尔·罗威尔（1855—1916）通过天文望远镜观测了火星表面，这大大激发了人们对火星上可能存在智慧生命的兴趣。遗憾的是，罗威尔所看到的火星运河，只不过是由于心理因素将模糊图像中的小点连起来而造成的错觉，而他关于火星文明的想法也都仅仅是幻想。随着天文学家更多地开始使用诸如行星光谱分析等更精细的观测技术，大家越来越清晰地认识到，太阳系中许多行星上的物理条件可能并不利于生命存活。局势开始不可避免地向相反的方向发展——到20世纪中叶几乎没有天文学家对行星感兴趣了。这一旧时的好奇心直到太空时代到来才被重新点燃。

尽管天体生物学研究的基本问题有悠久的历史，但是"天体生物学"这一专业术语一直都是偶尔现身，直到20世纪90年代才开始变得常见。1941年，纽约市立大学布鲁克林学院的哲学家劳伦斯·拉弗勒的一篇题为《天体生物学》的短文对该词的解释比现代的诠释狭隘一些，仅仅考虑了地球之外的生命。1955年，加利福尼亚大学伯克利分校的天文学家奥托·斯特鲁维也将"天体生物学"描述为对地外生命的搜寻。而俄罗斯天体物理学家加夫里尔·季科夫（1875—1960）和德国天文学家约阿希姆·赫尔曼则分别于1953年和1974年出版了名为《天体生物学》的书，涵盖了当时关于地外生命的流行观点。

现代意义上的"天体生物学"一词是由韦斯利·亨特里斯于1995年在位于华盛顿特区的NASA总部首次使用的。当时，NASA的科学家们认为，对生命进行从微生物到宇宙尺度的研究是理解宇宙中生命的必由之路。亨特里斯出于这一愿景而喜欢上**天体生物学**一词，并最终敲定这一术语。

实际上，天体生物学确实是**太空生物学**的重塑与扩展，而太空生物学研究可以追溯到几十年前。1960年，乔舒亚·莱德伯格（1925—2008）首次使用了这个词，指"地球之外生命的演化"。莱德伯格——因在细菌遗传学方面的发现而荣获诺贝尔奖——认为寻找生命是太空探索必不可少的一部分。随后，从20世纪60年代开始，NASA遵循了莱德伯格的建议，为太空生物学研究提供资金支持。但是，批评太空生物学的声音也随之而出现。例如，1964年，哈佛大学的生物学家乔治·盖洛德·辛普森曾讥讽道："这门所谓的'科学'甚至尚未证实其研究对象是存在的！"

现代的天体生物学则无须担忧如何回应辛普森的指责了，因为它的核心内容涵盖了对地球上生命的起源与演化的研究。天体生物学同样强调作为生命存在环境的行星的起源与演化，因此相比于太空生物学，天体生物学更加明确地涉及了天文学家们的研究。

太空生物学并不是唯一一个与天体生物学相近的术语。自1982年开始，天文学家已经正式使用**生命天文学**这一术语，指天文学研究中涉及地外生命搜寻的部分。更早些时候，德斯蒙德·J. 贝尔纳（1901—1971，著名爱尔兰裔英国物理化学家）曾青睐**宇宙生物学**。不过，它们都未曾被广泛使用。

何为生命？

天体生物学提出了如何定义生命的难题。换言之，我们要在地球之外寻找的究竟是什么？一个常见的办法是列举出生命的特征：繁殖、生长、通过新陈代谢来利用能量、对环境的应激、演化性适应，以及细胞层次上与解剖学上的有序结构。但是出于以下原因，这种定义生命的方式并不能令人满意。首先，枚举只是描述了生命是怎样的，而不能说明生命是什么。其次，这些特征大部分并不是生命所特有的。生命有结构秩序，例如细胞，但是盐晶体也是有序结构。我的一些朋友没有孩子，但是他们毋庸置疑是活的——我想这对于不生育的狮虎兽也是一样的道理。生长发育适用于生物体，但同样适用于蔓延的火。① 所有的生命都通过新陈代谢获得能量，不过我的车也一样。生命对周围环境有所响应，然而水银温度计也可以。

还有一些科学家试图用**热力学**，即热和能量及它们与物质的关系来定义生命。他们认为，生命的本质是通过由新陈代谢的废物和热量产生的**熵**来维持一种稳定结构，比如细胞和遗传物质。

在这里需要对熵这个术语做一些澄清。一些不考究的教师会用一个简单但具有误导性的词语"混乱度"来称呼它。熵并不是"混乱度"，而是对粒子间能量分布的精准度量——无论是原子还是分子。能量在空间上是分散的，因此一同运动的一组粒子的能量，也被称为**相干**能量，是会耗散的。因此，一个弹跳的球最终会因摩擦导致其相干能量转化为分子与原子的不相干热运动而最终停下来。相对地，一个静止的球永远不会像活物那样自发地开始弹跳，因为即使下面的地板中存在足够的热能，这个能量也是不可用的，并分散在地板原子的随机振动中。这

---

① "生长发育"的英文为growth and development，它有增大、扩展之义。——译注（本书页下注均为译注，不再一一标明）

种现象受控于热力学第二定律,即宇宙中的熵永远不会减少。熵增(或能量的分散)不改变能量的总量,却影响能量的品质。高品质的能量集中而不分散,就像聚集在一桶油中的原子核,或者像频率高、波长短的光子一样。这种光子包括可以晒伤人的紫外光和为植物生命活动供能的可见光。在物理学中,这种高品质的能量有着较低的熵。

将熵与生命联系起来的物理学家中,最著名的莫过于诺贝尔奖得主欧文·薛定谔(1887—1961)。他在《生命是什么》中曾谈道,一个生命体"倾向于接近熵最大的危险状态,也就是死亡。它只能从环境中不断获取负熵来远离这种状态,即活着……事实上,就高等动物而言,我们非常了解它们何以保持生存的秩序,即进食那些复杂程度不一但物质极度有序的有机物。在被利用完之后这些物质将回到充分分解的形式"。遗憾的是,薛定谔引入了科学中并不存在的概念"负熵"来描述食物的有序结构。此外,在一些生物体的生长过程中,熵的增加主要来自产热而非食物降解成代谢废物。被认为是20世纪最伟大化学家的莱纳斯·鲍林(1901—1994)曾直言不讳地评价:"[对于我们对生命的理解,薛定谔]没有做出丝毫贡献……他将'负熵'的概念与生命相关联的说法反而起了消极作用。"

尽管如此,宇宙中不断增长的熵的一个不寻常的副产品就是演化出有序、低熵结构,比如说生命体。事实上,最有效的熵增过程是由所谓**耗散结构**实现的。它是一个由大量**耗散**能量的粒子形成的相干**结构**。一个简单的例子是沸水中的对流现象,即热水的抬升伴随着边缘冷却水的下沉。这种对流单体有助于能量分散,从而比没有它时更有效地增加熵。所有活的生命体都是复杂的耗散结构。然而,到目前为止,使用热力学来定义生命的尝试都无法明确地区分生命与非生命。比如,作家埃里克·施耐德曾把生命定义为"通过产生环境熵以维持局部有序的、远离平衡状态的耗散结构"。但是,一团火也能满足这个定义。

抛开鲍林的批评不谈,薛定谔正确地指出,有机体必须运行一种类似于计算机程序的机制,这就是我们现在所称的**基因组**。确实,无论是何处的生命都可能必须拥有一套基因组。这里所说的**基因组**是指会有微小复制错误的遗传蓝图,使得生物体能够从其祖先演化而来并决定了生命的其他特征,例如新陈代谢。对个体特征选择导致的连续世代种群所发生的变化,即**演化**,也发挥了非常重要的作用,因为它是唯一能解释生物多样性和上述诸多生命特征之由来的过程。在达尔文的自然选择理论中,种群中个体的遗传变异意味着一些个体的适应能力强于其他个体,从而在繁殖上更加成功。自然选择会偏好能留下更多后代的基因,因此生物谱系不断积累遗传适应性。

考虑到演化的中心地位,天体生物学家通常把生命定义为"可以自我维持的、能够进行达尔文演化的化学系统"。很可惜,这个定义并不能帮助我们设计实验来寻找生命。难道我们一定要等到演化发生后才能检测出生命吗?一个更好的定义要用到过去时:"生命是可以自我维持的、包含基因组的化学系统,并且**已经**通过演化获得了其现有特性。"迄今为止,所有搭载在航天器上的生命探测仪器都没有设计成用来探测潜在地外生命的基因组成。例如,20世纪70年代,NASA曾发射"海盗号"着陆器寻找火星生命,主要的设计目标是在火星土壤中识别类似于地球微生物的新陈代谢现象（详见第六章）。

哲学家卡罗尔·克莱兰和科学家克里斯托弗·希巴曾经指出，我们现在定义生命的尝试很像17世纪一些科学家试图定义水。那时，水被认为是无色无味的液体，在一定温度下沸腾和结冰。没有原子理论，没人知道水是一些分子的集合，每个水分子都由两个氢原子和一个氧原子组成。这样想来，也许我们目前还缺乏定义生命所需要的生命系统理论。

我们在定义生命的过程中所遇到的许多问题都可以归结为：我们只有地球生命这一个例子。所有地球上的生物体都以核酸作为遗传物质，用蛋白质调控生化反应速率，依靠相同的含磷分子储存能量。无论是一个细菌还是一头蓝鲸，它们的基础生物化学组成都是一样的。因此，我们很难分清地球生命的哪些属性是其特有的，哪些又是"生命"所普遍适用的。如果我们真的在地球之外发现了生命，那天体生物学可以帮助我们解开这个谜题。

生命的最基本必需品

虽然尚没有对生命的完美定义，但我们有理由认为，地球生命的生物化学组成中的某些常见原子可能也会被地外生命利用。这种认识可以帮助我们识别地球之外的潜在生命。地球生命的主要构成元素是碳、氮和氢，同时，生物化学反应在液态水中发生。天体生物学界达成了一个广泛共识：其他地方的生命很可能也是碳基的，而一个有液态水的星球至少对"我们所认知的生命"是有利的。这样的推论源于我们意识到，生命是由一套有限的工具，即元素周期表构筑的，这在整个宇宙中都是相同的。

事实上，碳是唯一能够形成像DNA（脱氧核糖核酸）这样由数十亿个原子组成的长化合物的元素。由此可知，似乎只有碳基地外生命才能拥有与地球生命相当的复杂基因组。碳也拥有一些其他特性，使它的化合物具有独特的化学性质，足以延伸出一整个有机化学学科。碳的特殊性质包括：能与自身形成单键、双键和三键，并且能与其他很多元素成键。碳也可以通过将六元环连接起来，从而构建复杂的三维结构。

生命需要起源和繁衍，因此构成生命的主要原子可能也是在宇宙中大量存在的原子。碳在宇宙元素丰度中排行第四，仅次于氢、氦和氧。事实上，天文学家已经在太空中发现了许多非生物成因的有机分子。这些宇宙的"免费赠品"可能为生命起源提供了前体物质（详见第三章）。例如，以质量计，星际尘埃中约30%都是有机物，而我们太阳系中所谓的碳质球粒陨石和行星际尘埃颗粒，分别含有2%和35%的有机碳。

虽然硅在宇宙中的丰度只有碳的不到十分之一，但鉴于硅和碳具有类似的化学性质，也有人主张，硅可能支持形成一种非碳基分子的地外生物化学体系。可是，至少在水中，硅的化合物往往是不稳定的，且容易形成溶解度很低的固态硅氧化物。在常见的行星温度条件下，二氧化碳是气体，溶解在水中可以为生物体提供足量的碳源。与其相反，二氧化硅往往是溶解度很低的固体，例如石英。此外，硅氧键、硅氢键比较强，但碳氧键、碳氢键与碳碳键的强度接近，这使得碳基化合物可以进行交换和改变反应。硅氢键在水中很容易受到破坏，因此硅基分子需要低温环境来减缓它的水解反应。这种环境包括远离恒星的冰质行星上的液氮海洋。目前，这些对于硅基生命的探讨还仅仅停留在纯粹的推测。

不管怎样，稳定的介质对于新陈代谢和基因复制等生物化学过程一定是必需的。在地球上，这样的介质就是液态水。对于地外生命，这种介质可以是另一种液体或是某种稠密气体，只要它在所处环境中不容易变成固体即可。然而，水有一些独特性质。例如，在常压下，水会在100℃以下变成液体，而不像它的孪生子、散发着臭味的硫化氢，只有在-61℃时才会凝结成难闻的液体。液态水的稳定性主要归功于，水分子中的氧原子带轻微的负电荷，与其他水分子中带轻微正电荷的氢原子形成了较强的氢键，而硫化氢分子间的氢键更弱一些。水分子间的氢键强于水分子与油性物质分子间的氢键。正因如此，油与水分离，使细胞膜得以形成，为基因和新陈代谢过程提供稳定安全的环境。

水的另一个不同寻常的性质是，固态（冰）的密度小于液态。当水结冰时，分子排列成有原子尺度空洞的环状结构。如果冰比水更致密，冰将在湖和海中更冷的底部积累，使其隔热且保持冰冻的状态。大海将从底部开始向上结冰而变得不适合生命存在。这是因为当冰积累至海面时，阳光会被冰面反射，导致温度持续下降、冰继续积累。逐渐冻结可能是其他液体海洋，如氨海洋的悲惨命运。在一个标准大气压下，氨在-78℃到-33℃这个区间内呈液态。但任何氨海洋都倾向于从底部向上凝固，而不像水海洋那样即使在低温下形成冰盖也能在内部保持液态。

我们可以在地球上任何有水的地方（除了灭菌设备中）找到微生物，因此"我们所认知的生命"既以水为基础，又以碳为基础。这样一来，对于太阳系探索来说，寻找液态水或它们过去存在的痕迹是行星探测器，比如火星探测器的重要目标。尽管如此，仍然可以设想用有机溶剂作为水的替代品，这对土星最大的卫星土卫六来说可能很重要（详见第六章）。

对地球生命的另一个观察结果是，就质量而言，仅六种非金属元素——碳、氢、氮、氧、磷和硫——就构成了99%的生物体组成物质。这些元素通常被缩写为"CHNOPS"，但我更喜欢说"SPONCH"，因为更方便读出来。生命体的最主要成分水由氢和氧构成，遗传物质中的核酸以及糖类主要由碳、氢和氧构成，而蛋白质的主要成分是碳、氢、氮和硫，磷则是核酸和储能分子所必需的。因此，探测生命可利用形式中的SPONCH元素是行星空间探测器寻找类地生命的另一个可行的目标。

地外生命探索的重要性

你可能会想，发现简单的、类似微生物的地外生命是否真的有意义。但如果我们能找到一例起源于地外环境的生命，这就证明生命不是一个局限于地球之上的奇迹。我们便不再孤独。即使是在火星上或者木卫二海洋中发现最简单的微生物，也会极大提高在银河系中其他天体上存在生命的可能性，因为这个发现将表明即使在一个太阳系中也可以发生生命的两次起源。目前，我们还没有任何令人信服的证据能证明地球以外存在生命。不过，如果我们保持一个开放的心态，我认为太阳系中还有至少九个地外宜居天体（详见第六章）。太阳系的天体生物学还远没有定论。

寻找地外生命的第二个重要原因就要回到如何定义生命的

难题上了。行星科学家卡尔·萨根（1934—1996）在他的《宇宙的连接》(*Cosmic Connection*, 1973)中曾指出，"迄今为止，从行星探测中获益最多的科学是生物学"。探索地外生命的深远意义，不仅在于能鉴别所有生命共有的复杂特征，而且有助于阐明生命如何起源这一悬而未决的难题。

第二章

从星尘到行星,生命之所栖

> "如果你真想从头开始制作一个苹果派,那么你得先创造宇宙。"
>
> 卡尔·萨根(1980)

宇宙诞生时,各处温度都奇高无比,连原子都难以稳定存在,更别提结合成复杂的生物分子了。138亿年前的**大爆炸**后,高温、致密的宇宙逐渐膨胀与冷却下来,生命才得以存在。这期间,原子、星系、恒星、行星与生命次第登场。现在我们来看看这一过程如何产生了生命的家园——地球。

我们从现在的宇宙结构开始,这能为了解它的历史提供一些线索。现在,想象一次前往可观测宇宙边缘的旅程,我们将以光速——300 000千米每秒——前进,到达384 000千米之外的月球只需1.3秒。下面是一个直径2.7毫米的地球和相应比例的月球的示意图:

地球　　　　　　　　　　　　　　　　　　　月球

这个尺度下的比较是非常容易理解的。但是当我们意识到在相同的比例尺下,太阳的直径是30厘米并且距离上图的地月系统有30米时,这就开始有些挑战性了!在相同的比例尺下,离我们最近的恒星,半人马座的比邻星的直径将只有约4厘米,比太阳要小。但是,为了保持相同的比例尺,我们得把它放在距这本书8 650千米之外的地方,这大致相当于从旧金山到伦敦的飞行距离。

继续我们的光速旅程,我们将花费8.3分钟从地球飞临太阳,接着花4个多小时飞完至海王星的平均轨道距离,它位于八大行星的最外围。4.2年后,我们将来到半人马座的比邻星。为了更好地比较这些遥远的距离,我们将以光速运动一年的距离定义为1**光年**,约等于9.5万亿千米。因此,比邻星与地球的距离是4.2光年。

圆盘形的银河系直径达10万光年之巨,多达3 000亿颗恒星集中在它的旋臂中,而太阳和比邻星只不过是这3 000亿颗中的两颗。星系往往包含数百万到数万亿颗恒星,银河系只能算中等大小。太阳系坐落于猎户座旋臂,处在距银河系中心约三分之二半径远的地方。与其他那些恒星分布得更密集的旋臂相比,猎户座旋臂显得平凡而暗淡。但是太阳系在银河系旋臂中的位置可能对地球生命的诞生至关重要。地球可能因此避免了某些灾难事件,比如邻近恒星的爆炸。要是这样,在星系中可能会存在一个适合生命的特定区域,被称为"银河系宜居带",我们将在第七章详细讨论。

在更大的范围内，我们可观测到的宇宙中有超过1 000亿个星系，组成不同的星系群和超星系团。以银河系和离我们最近的螺旋星系仙女座星系（大约250万光年）中点为中心，在直径约1 000万光年的范围内，大约存在50个星系，一同组成了**本星系群**。按照这个规律，这个星系团及100多个周围的星系团组成了一个直径为1.1亿光年的**室女座超星系团**。在数十亿光年尺度的星图上，我们将看到由无数零散的小点所连接成的各种各样的纤维状结构，其中的每一个小点都是一个星系。而在三维空间中，人类已知的最大宇宙结构——宇宙纤维状结构——连接成一张有许多巨大孔隙的网。整个光怪陆离的结构看起来就像由一只疯狂的星际蜘蛛织就。

可观测宇宙有多大？如果宇宙空间没有膨胀，最远的距离将是138亿光年，因为大爆炸发生于138亿年前。但是宇宙空间已经发生了膨胀，所以可观测宇宙的实际大小约有470亿光年。如此浩淼的空间是天体生物学研究值得考虑的因素，因为对银河系内恒星周围行星的调查表明，平均而言，每颗恒星至少拥有一颗行星。其中一部分行星应该是宜居的，可能至少有1%。因此，生命的潜在宜居天体数量可能超过10万亿亿。

宇宙的结构要追溯到大爆炸时。20世纪20年代，美国天文学家埃德温·哈勃通过望远镜观测发现，由于空间膨胀，星系正在大规模地远离彼此。因此，在很久之前，所有物质应当是"揉作一团"且温度极高的。这种思考指引我们找到了支持"大爆炸"假说的强有力证据。如今，大爆炸的余晖——**宇宙微波背景辐射**——弥漫在整个宇宙之中。如果"大爆炸"假说是正确的，物理学理论表明，在早期宇宙冷却之前，它一定是一团不透光的火球,由电子及质子(分别是构成原子的负电性和正电性基本粒子)、光量子(简称光子)和一些聚变质子组成。在大爆炸后约38万年,宇宙开始冷却,温度降到电子能够与质子或一组质子结合形成最小的两种原子:氢原子和氦原子。从这一刻起,宇宙对光来说变得"透明"了。而在此之前,光子会与宇宙中的自由电子碰撞而发生散射,原理类似于光线在雾中微小的水滴上弹射导致雾不透光。自从变得可以透光以来,宇宙膨胀了约1 000倍,大爆炸残余光子的波长也因此拉伸了相同倍数,从红光变成微波。令人惊奇的是,给一台老式电视机或收音机调台时,你仍然能听到由大爆炸剩余辐射产生的静电发出的沙沙声,尽管只有1%或更低的水平。事实上,1964年人们正是从大型无线电天线的信号噪声中发现了微波背景辐射。最初,人们将噪声归咎于天线上的鸽子粪便。但是,在清理了粪便和除掉了那些不幸的鸽子后,人们才发现真正的罪魁祸首是宇宙的开端。

大爆炸发生数亿年后,星系才开始出现。一些地方有比平均水平略多的物质,从而产生了更大的引力。物质聚集产生了星系,而在星系内部更小的尺度上,气体云在自身引力下碰撞坍塌。随着气体云的收缩,气体分子间的碰撞导致星云内部升温,最终形成一个炽热明亮的气体球——恒星。

天体生物学中使用"天体"的部分原因是,生命所利用的所有原子,除了氢,都创造于恒星内部。第一批恒星应该只由在大爆炸中制造的元素组成:氢(占总质量的四分之三)、氦(占总质量的约四分之一),以及少量的锂。水中的氧、蛋白质中的氮或是所有有机分子中都含有的碳,这些元素都不是一开始就存在的,而是恒星最终把它们制造出来的。

要了解恒星如何制造元素,可以想想太阳是怎样发光的。在太阳内部,巨大的热量将每个原子都剥离成了它的基本组成部分,即带正电的原子核和带负电的电子。太阳核心的温度高达1 600万摄氏度,足以使四个氢原子核发生聚变,产生一个氦原子核。这一核反应会伴随光子的释放。每一个光子都要经过100万年的漫长旅程才能从太阳内部到达外部太空。之所以需要这么长的时间,是因为每一个光子在遇到物质时都会被不断地吸收和发射出来,这个过程也存在能量损失。开始时,光子是高能的伽马射线,而当它从太阳内部逃逸出来时,整体上变成了能量较低的可见光。20世纪50年代,物理学家们用氢弹再现了太阳内部的核反应。事实上,因为被外部物质的巨大质量所包裹,像太阳这样的恒星,核心类似于不会爆炸的氢弹。

太阳核心的氢聚变不会永远进行下去,但它最终会摧毁地球上的生命(这里部分地回答了天体生物学的经典问题"生命的未来会怎么样?")。在大多数恒星的核心,氦"灰烬"的积累会导致温度持续降低,以至于无法继续诱发氢聚变。此时,恒星在自身引力作用下收缩,温度随之升高,直到在核心周围的壳层引发氢聚变。聚变释放的能量会使恒星外层膨胀、冷却和变红。这就是**红巨星**的形成过程。金牛座中最亮的恒星毕宿五就是一颗红巨星。太阳最终也会变成一颗红巨星,将在75亿年后膨胀200倍,到时候可能会吞没地球。氦的进一步形成和积累将会挤压红巨星的核心,使其温度达到1亿到2亿摄氏度,足以让氦发生聚变产生碳和氧。相应地,太阳内部将会形成一个发生氢聚变的壳层,包裹着由碳和氧"灰烬"组成的核心。对于4到8倍于太阳质量的恒星,它们甚至会发生碳和氧的进一步聚变,形成更重的元素，包括氖和镁。

一般来说，在类太阳恒星垂死挣扎的过程中，它的外层会脱落到太空中。这些发光气体壳层被称为**行星状星云**，因为在低倍率望远镜下它们看起来就像行星一样，但是跟行星毫无关系。恒星的残骸冷却下来形成**白矮星**，它们的大小和地球相仿但密度巨大。理论上，白矮星会在几百亿到几千亿年后停止发光，成为**黑矮星**，但宇宙还没有那么老。

质量超过太阳8倍的恒星最终会爆发成**超新星**。目前，太阳核心的氢聚变阶段已持续46亿年，到达了100亿年寿命的半途，但是大质量恒星往往经过不到6 000万年就会变成**红超巨星**，例如猎户座中的参宿四。在这些大质量恒星内部，环绕核心的聚变壳层会产生氖、镁、硅、铁等元素。铁通常是所生成元素中最重的，但有些更重的元素可以通过自由中子加入已有原子核而产生。（中子是不带电荷的粒子，通常存在于原子核中。）当这些恒星的燃料燃尽时，核心压力增大，以至于带负电荷的电子与铁原子核中带正电荷的质子结合。电荷相互抵消，从而产生不带电的中子，其结果就是恒星核收缩成一个半径为12千米的、只由中子构成的"巨型原子核"。收缩过程会伴随恒星的其余部分向致密的中子核坍缩以及剧烈的回弹，形成超级明亮的超新星。

超新星产生并向宇宙中提供了比铁重的元素。超新星爆发几秒后，其外层被加热到了难以想象的100亿摄氏度，内层发生原子核的破碎，为制造更重元素的反应提供了充足的中子[①]。这种宇宙炼金术制造出金、银、铂等重而珍贵的元素。超新星爆

---

① 作者将此处更新为："为制造新的、包括比铁更重的元素的反应提供了充足的中子。"

发的重要生命效应是,它喷出的物质构成了新一代恒星与生命的居所——行星——的基础。如果恒星的质量达到太阳的几十倍,那么尽管仍然会有超新星爆发,但它核心的坍缩会产生一个**黑洞**——一种质量极大的天体,没有任何东西能逃脱它的引力,包括光。

通常认为,恒星周围的行星孕育出生命的最佳时间是,恒星处于核心发生氢聚变成氦的反应的**主序**阶段。在天文学最著名的**赫罗图**(图1;Hertzsprung-Russell图,得名于它的两位发明者:赫茨普龙与罗素)上,"主序"指的是从左上到右下的对角线

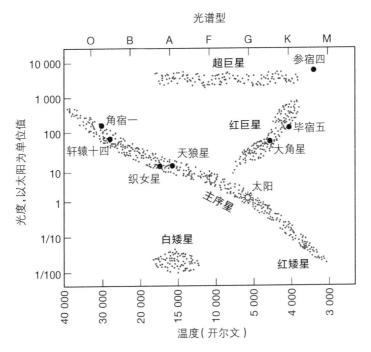

图1 赫罗图,其中标注了太阳与邻近恒星。温度以开尔文为单位,即摄氏温度加273;在开尔文温标的定义中,在0 K或"绝对零度",分子运动处于完全停止的状态

涵盖的区间。这幅图展示出了恒星亮度与表面温度的关系。这里需要指出的是，天文学家所说的"表面"并不是指某个坚硬的表面，而是指恒星大气层中发射大部分光的层面，也是我们目前能看到的最深的地方。

奇怪的是，赫罗图中温度轴是反过来的，从高温到低温。这是为了与使用字母O、B、A、F、G、K和M来代表从炽热的蓝白色恒星到较冷的红色恒星的颜色编码保持一致。这些字母代表了恒星的**光谱型**。一代又一代的天文学学生都利用"哦，做个好女孩/男孩，吻我！"（Oh Be A Fine Girl/Guy Kiss Me!）的口诀来记忆恒星光谱型。这些字母本身并不代表任何东西，最早是19世纪的天文学家用来代表光谱型的，与我们没什么关系。

在主序中，质量最大的恒星在左上角，质量最小的恒星在右下角。无论质量如何，主序星都被叫作矮星，比如太阳就是一颗G型矮星。对于冷红矮星来说，主序星阶段可持续超过500亿年。

恒星"生存"与"死亡"的方式对天体生物学的影响是十分广泛的。我们的太阳是一颗正值中年的主序星，为地球上大多数生命依赖的光合作用提供稳定的能量。正如一开始所提到的，构成生命的原子产生于红巨星和超巨星。此外，氧、硅、镁和铁等元素最初都是由整数个氢原子聚变而成的，并且核聚变过程产生了大量的这些元素。这对于生命来说也是非常重要的，因为正是这些原子构成了岩石。也就是说，包括我们生活的地球在内的岩质行星都是恒星光芒背后的物理过程的自然产物。我们目前还知道太阳至少是第二代恒星，因为地球上存在像金这样的超新星元素。既然在132亿岁高寿的银河系中太阳只存在了46亿年，那么在地球生命诞生前，自然已经发生过许多次恒星的诞生与毁灭。

更早的恒星支持行星与生命,甚至智慧生命吗?它们发生了什么?这引出了一个问题:我们的行星是如何形成的?

寻生命之所栖:行星从何而来?

人们很早就开始思考太阳系的起源。1755年,伊曼努尔·康德提出,太阳系是由空间中弥散的云团聚合形成的。随后,在1796年,数学家皮埃尔-西蒙,也就是我们所熟知的拉普拉斯侯爵,详细阐述了这一观点。康德-拉普拉斯的基本观点就是后来为人熟知的"星云"假说(nebular hypothesis),在希腊语中nebula意指"云"。

该假说的出发点是,星云的某些部分比其他部分密度略大,并通过引力吸引物质。星云可能初始时慢慢转动,因此收缩时会转得更快,就像滑冰时收回手臂一样。气体与尘埃的随机运动会阻碍物质沿旋转轴相互吸引,而在旋转平面内汇聚的物质也会因旋转而受阻。最终,星云变平成为星盘,太阳在中心形成,而星盘平面上稀疏的物质合并成行星。这个假说与我们目前观测到的行星只占太阳质量的0.1%的事实相符。由于太阳系中各行星形成自同一个星盘,因此它们处在同一平面,并且沿同一方向绕太阳运动,这也与我们观测到的情况一致。

近几十年来,一些同位素证据表明,附近的一次超新星爆发产生的冲击波触发了上述星云的坍缩过程。同位素是原子核中含有相同数量的质子但不同数量的中子的原子。同位素的希腊语词根isos topos意为"同等的排位",指的是在元素周期表中占相同的位置。有时某个同位素的原子核太大而难以稳定存在,会通过放射性衰变而分裂。例如,不稳定的铝-26(原子核中有

26个粒子,包括13个质子和13个中子)会衰变为稳定的镁-26(包括12个质子和14个中子)。在任何含有铝-26原子的样本中,一半的铝-26原子转变成镁-26所需的时间,也被称为铝-26的**半衰期**,是70万年。一些古老陨石中存在镁-26,并且一定是由铝-26在太阳系形成时快速衰变产生的。学者认为,这表明太阳系形成时附近发生了一次超新星爆发事件,因为大质量恒星会制造出铝-26并在爆发事件中将它们散播出去。

然而,在2012年,新的陨石研究数据显示,一种只在超新星中形成的同位素铁-60含量过低,不能解释上述的"邻近超新星爆发"假说。为了解释陨石中大量的铝-26,人们又认为可能是一颗邻近的大质量恒星(质量为太阳的20多倍),被称为沃尔夫-拉叶星,剥离了它的外层,将铝-26散播到形成太阳系的原始星云中。在最初的几百万年,铝-26的衰变是太阳星云的主要热源。铝-26衰变释放的热量能融化最早的岩石物质中的冰,使水逐渐转化为水合矿物并安全地保存下来,而不会像冰那样升华。如果铝-26不存在,地球上可能不会有富水矿物,也就不会形成海洋和生命。

"星云"假说可以解释不同类型行星的广泛分布特征。内太阳系的行星(包括水星、金星、地球和火星)相对较小且均为岩质行星,而外太阳系的行星(包括木星、土星、天王星和海王星)则是巨大的。当星盘形成时,被吸向中心的物质获得动能,因此形成太阳的星盘中心变得极为炽热。这形成了由较热中心向较冷外围温度降低的梯度。在太阳之外,压力较低,意味着物质只能以固态或气态,而非液态存在。相较而言,星盘内侧又非常热,水蒸气这样的气体难以凝华成冰,但金属和岩石蒸气可以在星盘的几乎任何地方凝华。因此，靠近星盘内侧形成的行星，如地球，最终会形成富铁的核与包围它的岩质的幔。相反，水蒸气会自木星轨道内侧（"冰线"）往外凝华成冰，为构成更大的行星提供了更多物质。甲烷也可以在海王星轨道以外的天体表面凝华成固体。

巨行星被认为比岩质行星先形成。木星和土星很可能是在它们的原始岩质核心达到地球质量十倍左右的时候形成的，此时巨行星的核有足够的引力直接从原始星盘中吸引越来越多的氢气和氦气，直到它们轨道上的所有可用气体都被吸走。它们是主要由气体构成的巨大球体，所以我们称这样的行星为**气态巨行星**。这一过程发生在太阳形成后约 1 000 万年内。天王星和海王星更小一些，吸引了更大比例的冰质固体，因此我们称这些行星为**冰质巨行星**。相比于木星那样快速生长，内太阳系岩质行星的形成延续超过 1 亿到 2 亿年。星盘中的星子，即"行星的碎片"，合并成更大的、名为行星胚胎的岩质天体，其大小介于月球和火星之间。按照这个规律，若干个行星胚胎融合形成了金星和地球，少数形成了水星和火星。

尽管内太阳系行星主要从自己的位置上吸聚物质，但来自巨行星，特别是木星的引力会把一些星子推入内太阳系。例如，来自火星轨道之外的零散含水小行星很可能为早期地球带来了水，最终形成了我们现在的海洋和湖泊。也就是说，我们正在喝着来自小行星的水。计算机模拟显示，木星向轨道外喷出的富水物质比向内散播的更多，因此如果木星轨道没那么圆而且向轨道外喷出更多富水物质的话，地球可能最终无法形成海洋与生命。

我们还知道，行星不一定会一直待在自己的轨道上。例如，科学家们观测到一种质量类似于木星的系外行星，叫**热木星**，它们绕主恒星运行的轨道半径最多是日地距离的二分之一。但是，这种系外行星现在所处的位置太热了，无法支持原位的形成过程。事实证明，原始星云中与行星伴生的气体和尘埃拖尾，以及行星和其他天体间的引力都会导致行星的轨道迁移。因此，**行星迁移**使得传统的"星云"假说多了一丝不确定性。取决于具体的情况，行星迁移效应可能破坏抑或促进太阳系外行星的宜居性。在下一章中，我们将讨论，太阳系的巨行星也可能发生过一定的轨道迁移。

尽管如此，在20世纪80年代，人们通过望远镜观测到围绕年轻恒星的碎屑星盘，证实了"星云"假说的基本思想，即行星形成于原始星盘。事实上，在我们太阳系中仍残留着一些这种碎屑。冰质的彗星与岩石质的小行星都是未被融合的残片。小行星在撞击中产生的碎片偶尔会以陨石的形式落到地球表面。因此，陨石会为我们提供早期太阳系的关键数据。

地球与月球的年龄

也许从陨石中得到的最重要的信息是地球与太阳系的年龄。自18世纪以来，地质学家观察到厚厚的沉积岩层，已经开始怀疑地球乃至太阳系应该年龄很大了，但是一直缺乏证据。

第一个试图测定地球年龄的人是英国地质学家亚瑟·霍姆斯。他想到可以通过分析放射性铀产生的铅同位素来测定地球的年龄。铀-238可以自然衰变成一系列其他元素的不稳定同位素，直到成为稳定的铅同位素，铅-206。铀-238衰变成铅-206的半衰期为44.7亿年。另一种放射性同位素铀-235衰变成铅-207，半衰期为7.04亿年。因此，通过测量岩石中不同矿物颗粒里铅-207与铅-206的含量就可以确定岩石的年龄，因为尽管各矿物颗粒中铀的含量不同，但在给定时间内固定的衰变速率给它们添加了相同比例的铅-206与铅-207。1947年，霍姆斯利用这种方法分析了来自格陵兰岛的一块铅矿石，并估算出地球形成于34亿年前。

霍姆斯的估算存在两个问题。首先，他无法保证他能找到的最古老的岩石就是地球诞生时形成的。其次，为了计算精确的年龄，霍姆斯需要知道地球刚开始形成时的、没有经过后期铀衰变的、小的原始铅同位素比值，即所谓的"原始铅"。美国地球化学家克莱尔·帕特森意识到，可以通过分析陨石来避免第一个问题，因为陨石是地球形成后遗留的"建筑材料"。他也认识到，某些特定类型的陨石，如铁陨石，含有微乎其微的铀，因此它们的铅同位素比值可以用于计算原始铅。通过这种方法，帕特森在1953年比较准确地将地球的年龄确定为45亿年。得到这个结果后，他十分兴奋，甚至担心自己是不是心脏病发作了，他母亲不得不把他送到医院。

自那时起，放射性同位素定年技术不断改进，为我们提供了地球形成时期重要事件的时间表。陨石中最古老的矿物颗粒表明，太阳系形成于45.7亿年前。地球紧接着在45.4亿年前形成。随后，在约45亿年前，地球被一颗火星大小的天体撞击，这颗天体以生下月亮女神塞勒涅的希腊女神忒亚命名。根据这个**"大碰撞"假说**，大碰撞抛出的碎片进入围绕地球的轨道，最终聚合形成了月球。

月球对天体生物学很重要，因为它的引力维持了地球自转轴倾角的稳定，有助于地球维持相对平稳的气候。目前地轴倾角为23.5°，但如果它剧烈变化的话，气候也会随之剧烈波动。比如说，在90°的倾角下，地球会完全倒向一侧，赤道将会季节性地结冰。计算机模拟显示如果没有月球，地轴倾角将在数百万年的尺度上混乱地变化，并且变化范围巨大，可能从0°到超过50°。这种巨变引发的剧烈气候波动可能对微生物来说是可以承受的，但对于高等动物和文明，比如人类来说将会是很大的挑战。

沿着我们的天文足迹，现在已经为生命创造了居所，我们的行星。这个过程既有偶然性，例如生命必需的水能否被送到地球，也有各种控制条件，这些条件最终决定了类地行星在哪里形成。那么，在地球形成后，生命又是如何以及何时起源的呢？

第三章

生命和宜居环境的起源

早期地球

我们对于地球历史上最早的时期知之甚少,以至于还没给它起一个正式的名称。这段漫长的时期一般被称为冥古宙,始于月亮形成的时间,大概是45亿年前。它结束的时间现在还没有定论,一般认为是38亿或40亿年前。[①]生命很可能起源于冥古宙,但是由于这个时期的沉积岩记录缺失,我们目前并没找到支持这个观点的证据。沉积岩由一层层从水中或者从大气中沉降下来的沉积颗粒组成,因此最好地保存了生物或环境留下的痕迹。所以,我们只能依靠稀少的数据和理论限制来构建对冥古宙的认识。

理论研究表明,形成月球的大碰撞所产生的热量会把岩石

---

① 基于2022年国际地层委员会对冥古宙、太古宙的正式定义,作者将此处更新为:"我们对于地球历史上最早的时期知之甚少,不久前名称才得到正式承认。这一时期被称为冥古宙,自45.67亿年前月球形成开始,到40亿年前结束。"

变为气体。由岩石升华而成的大气维持了几千年,然后凝结并降落到熔融岩浆的表面,这层物质最终会凝固成一个固体外壳。在随后的几百万年里,大气主要由极其稠密的水蒸气组成,最终水蒸气凝结形成了原始海洋。

大陆什么时候开始形成?我们从地球历史上最初的5亿年遗留下来的、一些直径不足0.5毫米的微小矿物颗粒中找到了线索。这种矿物叫**锆石**,是硅酸锆形成的晶体,化学式为$ZrSiO_4$。锆石坚硬稳定,即使所在的岩石已被侵蚀消失,它仍可保留下来。在位于澳大利亚西部的佩斯市之北约1 000千米的杰克山上,人们在一些已经老化的砾石中发现了44亿至40亿年前的古老锆石。这些锆石中包含了一些石英微粒,石英即二氧化硅的一种晶体,化学式为SiO_2。这些石英可能来自一种富含二氧化硅的火成岩——花岗岩,它是现今大陆的主要组成部分。因此,这些锆石证据显示,大陆地壳已经存在了43亿年之久。同位素证据也支持这一结论。科学家发现一些古老锆石的氧-18/氧-16同位素比值比较高。这种同位素富集特征往往发生在火成岩在地表水的作用下风化成黏土矿物时,它伴随着黏土矿物的埋藏和变质作用,最终记录在早期的锆石中。

在冥古宙,地球可能被一些太阳系形成时期遗留的巨大碎片撞击过,不过这些碎片都没有形成月球时的那个撞击体大。为数不多的几次非常大的撞击所产生的能量,可以导致整个海洋或海洋上层几百米的水蒸发。在这种情况下,原始生命将不得不重新开始,或者可能只有那些生活在地下且耐高温的微生物幸存下来。事实上,这种高温灭活的撞击可能也解释了地球上所有生命的最后共同祖先的特征。遗传学将地球生命的共同祖先追溯到嗜热菌（详见第五章）——一种生活在炎热环境中的微生物。换句话说，DNA分析指出你的"曾曾曾……祖母"是一个嗜热菌，这里你得插入足够多的"曾"字。这可能是由于嗜热菌是大撞击的唯一幸存生物。

今天的地球并没有完全摆脱潜在的灾难性撞击。例如，喀戎星是外太阳系一个直径约230千米的类彗星物体，其轨道与土星轨道交会。在未来约1000万年之内的某一天，土星引力的轻推会将喀戎星抛向太阳或抛离太阳。在前一种情况下，喀戎星撞击地球的概率不到百万分之一。但万一发生的话，撞击产生的热量会把海面以下几百米深的海水变成水蒸气，地表至地下50米左右的生命也会被全部消灭。如果上述假说成立，即早期地球撞击筛选出嗜热菌成为现在生物的祖先，那么这些嗜热菌可能会在喀戎星撞击后再次成为下一轮生命的共同祖先。这似曾相识的一幕可谓演化史上的"既视感"（dé jà vu）。

在大约40亿至38亿年前，发生了冥古宙撞击事件的最后一次狂欢，被称为**晚期重轰炸**。月球上的陨石坑见证了这一大规模撞击。科学家对"阿波罗号"宇航员带回的岩石进行放射性同位素定年，结果表明，许多月球陨石坑都是在这短短的2亿年内产生的。①

来自法国尼斯的天文学家提出了解释晚期重轰炸的主流

---

① "晚期重轰炸"现在被许多人认为可能是样本偏差的产物，作者将此处更新为："在大约41亿至38亿年前，可能发生了冥古宙撞击事件的最后一次狂欢。这一系列假设的撞击事件被统称为**晚期重轰炸**。月球上的陨石坑提供了关键证据。科学家对'阿波罗号'宇航员带回的岩石进行放射性同位素定年，结果表明，许多月球陨石坑都是在这短短2亿年内产生的。然而，有人质疑这些岩石样本可能本来就来自约39亿年前一次单一的大规模撞击事件。这次大规模撞击事件也塑造了月球的雨海盆地，因此集中分布的定年结果可能只是一个人为的统计假象。"

假说，因此该假说被称为**尼斯模型**。[①]它基于一个惊人的想法，即木星和其他巨行星的轨道在冥古宙末期发生了迁移。计算表明，在太阳系形成后，相互的引力作用使土星和木星达到了一种被称为"共振"的状态，即木星每绕太阳转两圈，土星就绕太阳转一圈。周期性的规律排列增强了引力的作用，使土星和木星的轨道变得不那么圆。海王星和天王星的轨道也因此受到扰动，向外迁移，变得更扁。海王星甚至可能从天王星的轨道内部开始迁移，逐渐越过了天王星轨道。这两颗行星，尤其是更远的那颗，会散射一些小的冰质天体，有些朝内太阳系运动。与此同时，木星的引力和运动会将一部分小行星抛向内太阳系，而把其他的推向更远的轨道。在这段时间里，地球由于体积和引力更大，受到的撞击一定比月球更多。最终，在太阳系遭到巨大破坏之后，巨行星的轨道稳定下来。

生命的起源

生命究竟是如何诞生的还不得而知。它可能在地球上诞生，也可能由太空尘埃或陨石带来。后一种观点被称为**生物外来论**，但它并没有回答生命是如何起源的，而是把问题推到了其他地方。此外，如果生命来自其他恒星系，它们可能难以挨过漫长的星际旅行。由此，我们下面将主要讨论生命的地球起源。

目前，学界有一个广泛共识：生命起源之前应该有一段时期进行化学演化，也就是**前生命化学**。在这段时间里，相对简单的有机分子产生了更复杂的有机分子。这个想法可以追溯到19世纪。

[①] 作者将此处更新为："尽管如此，来自法国尼斯的天文学家提出了解释晚期重轰炸的一个假说，因此该假说被称为**尼斯模型**。"

1871年，查尔斯·达尔文在给植物学家约瑟夫·胡克的一封信中写道，这样的前生命化学反应可能发生在一个"温暖的小池塘"里：

> 人们常说，第一个生命诞生所需的所有条件如今都已经具备，也许过去也一直具备。我们可以假设（哦，多么不简单的假设）某个温暖的小池塘具备氨、磷酸盐，以及光、热、电等条件，形成了某种蛋白质化合物，然后这种化合物再进行更复杂的变化。放在现在，这种物质会被立即吞食或吸收，但在生物诞生之前，情况并不是这样的。

因此，达尔文得出结论，生命不太可能在现在的环境中起源，因为现有生物体一直在消耗环境中生命起源所需的化合物。但是在另一方面，在生命诞生之前，适合生命起源的化学条件可能普遍存在。

20世纪20年代，俄罗斯生物化学家亚历山大·奥帕林和英国生物学家约翰·伯顿·桑德森·霍尔丹都认识到，生命起源的原始环境是缺氧的。地球上富含氧气的大气是植物、藻类和细菌共同光合作用的产物。无氧的大气更适合前生命化学演化，因为氧气会将有机物氧化成二氧化碳，不利于复杂分子的形成。1927年，近代地质学家亚历山大·麦格雷戈指出，太古宙（距今38亿至25亿年）[①]的沉积岩表明古老大气中缺乏氧气。特别是，麦格雷戈观察到现存于津巴布韦的古老铁矿物是没有被氧化的，完全不像在现代沉积物中看到的大气氧气与含铁矿物

[①] 基于2022年国际地层委员会对太古宙的正式定义，应修改为"距今40亿到25亿年"。

反应生成的呈铁锈色的氧化铁。现今，麦格雷戈的推论得到了许多其他数据的支持（详见第四章）。

奥帕林和霍尔丹还提出，地球早期大气中的多种气体可能会被紫外线或闪电转化为有机分子。20世纪50年代，诺贝尔奖得主、化学家哈罗德·尤里的学生斯坦利·米勒在芝加哥大学通过设计实验验证了这种想法。米勒将氨、甲烷、氢和水蒸气充入一个烧瓶中，然后释放电火花来模拟闪电。米勒发现瓶底逐渐凝结出淡黄色的溶液。这些深色的残留物含有多种有机分子，包括构成蛋白质的多种氨基酸。当时，米勒和尤里的实验结果被媒体大肆宣扬，认为生命起源的奥秘快要揭开了。生命也被说成是起源于大气化学产生的"原始汤"中。1953年，当米勒写下他的研究结果时，大家还普遍认为（甚至可以追溯到奥帕林和霍尔丹）遗传物质是蛋白质。但是，在米勒的论文发表一个月前，DNA被确定为真正的遗传物质。随后，类似的实验也产生了含有由碳、氮和氢原子组成的六边形环的分子，这个环就是在DNA中发现的那种。

然而，地球化学家对米勒-尤里实验提出了质疑，他们认为地球早期的大气可能不像米勒假设的那样富含氢气等还原性气体。在长时间尺度上，火山供给了地球大气中的气体成分，但火山气体的最主要成分是水蒸气（H_2O）以及平均少于1%的氢气（H_2）。同样地，火山气体中的碳主要是二氧化碳（CO_2），而不是还原形式的甲烷（CH_4），且氮主要以氮气（N_2）的形式释放，而不是氨（NH_3）。

目前认为冥古宙的大气主要来自火山活动释放的气体，因此当时的大气应该主要由二氧化碳和氮气组成。当然，原始大气的组成也取决于你往前追溯了多久。在地球形成时，大型撞击体的升华可能会产生富含氢的大气，尤其是当撞击体中含有足够丰富的铁，从而可以通过化学反应来维持撞击爆炸中富含氢的气体时。地球形成初期，大气主要由水蒸气组成，还没能凝结形成原始海洋。在原始海洋形成后，撞击活动可能会产生短暂的还原性大气。考虑到冥古宙大气成分的各种不确定性，天体生物学家提出了除米勒和尤里设想的闪电合成途径以外的多种有机碳的来源。

其中一种说法是，有机碳来自太空。一些富含碳的陨石含有多种氨基酸、醇类和其他有机化合物。这些化合物可能为早期地球提供了前生命化学所需的原料。例如，1969年坠落在澳大利亚默奇森的默奇森陨石，含有超过14 000种不同的分子。此外，0.02至0.4毫米大小的微陨石颗粒现在每年为地球带来约3 000万千克的有机碳。微陨石颗粒不一定会在大气中发生燃烧，因此这些有机碳可以完好地落在地球表面。此外，彗星内部富含有机物质。如果这些有机物质能在彗星撞击地球的过程中幸存，它们可能在生命起源方面扮演着与陨石类似的角色。

另外一个可能的有机碳来源是深海的热液喷口。海底热水不断涌出的地方形成了热液喷口。新的海底形成于洋中脊，在那里地球的构造板块发生分离，岩浆从地幔中升起，填充了板块分离导致的空缺。海水通过洋中脊附近的裂缝下渗，被炽热的岩浆加热后上升，带走了水和岩石之间化学反应产生的多种物质，如氢气、硫化氢和溶解的铁。当酸性热液遇到冷的（2℃）海水时，它会形成一团含深色黄铁矿（一种常见的硫化铁矿物FeS_2，其中Fe是铁，S是硫）的沉淀物。这就是这些热液喷口被称为"黑烟囱"的原因。一种假说认为,生命起源于这些黄铁矿等含硫矿物的表面。然而,"黑烟囱"非常热,大约350℃,所以大多数支持深海生命起源的研究人员关注的是温度较低(约90℃)、呈碱性而非酸性,并且离洋中脊更远的热液系统。

在这些碱性热液喷口下,水和岩石之间的反应会释放出氢气,而氢气可以与二氧化碳结合,产生甲烷和更大的有机分子。在冥古宙,喷口处可以生长出具有微小空隙的多种矿物结构。这些空隙一般含有简单的有机分子,它们逐渐浓缩,从而可以产生更复杂的前生命分子。

此外,这些碱性喷口流出的热液比海水的碱性更强,此处因而产生了与细胞内外类似的pH值梯度。如果生命真的起源于这样的环境,这也许能解释为什么能量的产生与pH值梯度密切相关。

生命通过细胞膜上的pH值梯度驱动电流,从嵌在细胞膜上的微型电动机中产生能量。这些令人惊叹的分子机器是无法用语言来准确描述的,所以我建议读者上网搜索"三磷酸腺苷(ATP)合成酶演示"来体会。大体上,生物利用新陈代谢产生的能量在细胞膜两侧表现出不同的pH值,细胞膜的外部通常比内部酸性更强。这个差值代表的是**质子梯度**,因为pH值与溶液中带正电的氢离子(质子)的浓度成反比,pH值越低(酸性越强),代表氢离子浓度越大。质子梯度本质上就是一个电池。放电时,电流通过细胞膜中的分子涡轮机,同时生成储存能量的分子。这些ATP分子是所有细胞中的能量载体。例如,一个人体内通常有250克ATP,在体内ATP不断被使用和重新生成的过程中,每天消耗大约跟体重相当的ATP。

上述产生ATP的过程被称为**化学渗透**（chemiosmosis），"渗透"（osmosis）的意思是水穿过膜，因为质子在细胞膜中的运动与其类似。与我们熟悉的通过化学反应产生能量的观念相比，化学渗透是如此奇特，以至于当英国生物化学家彼得·米歇尔在1961年提出它时，大多数生物化学家都持反对态度。最终，米歇尔的假说被证实，他因此获得了1978年的诺贝尔奖。考虑到化学渗透作用在地球生命中是普遍存在的，并且它可能与生命起源有关，我们可以思考一个有意思的问题：离子梯度是否可能是宇宙中任何潜在生命的基础？

目前，我们尚未确定生命起源所需有机碳的来源到底是大气、太空，还是热液喷口。然而，无论来源是什么，有机化合物都必须富集在一起才能发生化学反应。这个过程可能发生于在干燥或者寒冷条件下形成的矿物表面薄膜上。在这种情况下，简单的有机物可以相遇，形成更复杂的分子。但是，这些化学系统还需要一些其他特征才能被认为是"有生命的"。其中一些主要的生命特征是新陈代谢、半透膜隔离，以及通过基因组实现遗传的自我复制。

关于最后一个生命特征，对于实验室条件来说，比生命起源本身更容易处理的是生命早期的一个假设阶段，即**RNA世界**。这个假说认为，核糖核酸（RNA）是比DNA更早的原始遗传物质。在现代细胞中，RNA承载着自DNA转录而来的一套指令，用其中的信息指导特定蛋白质的合成。DNA与RNA相似，但更复杂，所以先演化出来RNA是合乎逻辑的。此外，在20世纪80年代，人们发现一些被称为核糖酶的RNA分子可以充当催化剂。因此，我们似乎有理由认为RNA曾经催化自我复制以及较小分子的组装。在生命演化的下一阶段，RNA将开始合成蛋白质，其中一些蛋白质是比RNA更好的催化剂。更稳定的结构和更大的体积为DNA提供了更多的繁殖优势，最终使它替代了RNA。

生命起源的另一个关键特征是基因组被封装在细胞膜中。这为早期生命提供了两个好处：第一，细胞可以通过浓缩生物化学物质来提高反应速率；第二，为自我复制的分子提供演化优势。例如，一个产生有用蛋白质的RNA基因组可以将自己封闭在细胞内，从而有利于其后代的生存发展。

现在，所谓的"前细胞"结构已经可以在实验室中人工制造出来了。我们在厨房里也能经常看到这种结构：当你把油和水混合时，类似的球形结构会自发形成。这种类型的油膜或脂肪膜可能最终演化成包裹现代细胞的膜，并为前生命化学提供表面载体。

在原始细胞膜形成后，生命起源可能见证过如下后续事件。RNA留在了前细胞中，含RNA基因组的原始细胞逐渐演化。随后，现代细胞用DNA基因组取代了RNA基因组。在其中的某个阶段，也许是在RNA世界之后，新陈代谢也逐渐形成。然而，学界仍然在争论新陈代谢和基因复制哪一个先出现。此外，人们仍然不确定生命起源发生的位置、环境条件和演化步骤。这将是一个亟待取得重要进展的研究领域。

手性（或孤掌拍手的技巧）

到此，本书对于生命起源的描述还是不完整的，因为还没有提及生命的一个独特的生物化学特征，即对于具有两种镜像结构的生物分子来说，地球上的所有生命只使用了其中一种结构，而不是两种都用。镜像对称的特征被称为**手性**（chirality），它来自希腊语cheir（手），就好似一个人的双手是镜像对称、无法相互重叠的一样。如果你不相信左旋和右旋两种结构是真正不同的，试一下一整天都反穿鞋子！

如果生物分子的一个中心碳原子被四种不同的原子团包围，这个生物分子就具有了手性。在空间上，这四个不同的原子团会围绕中心碳原子自然地形成一个四面体结构，像一座有四个三角形面的金字塔。如果我们把四面体一个顶点的原子团用"1"来编号，那么另外三个顶点可以按顺时针或逆时针顺序分别编号，如下图所示：

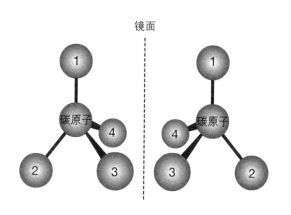

上图中的两个分子就是"手性"对称的，因为一个分子不能完全叠加在另一个分子上。这样的两个分子被称为**对映异构体**（enantiomers），这个单词源自希腊语enantios morphe，意为相反的形状。这里，我们考虑一个特定的有机对映体——丙氨酸。对应上图，丙氨酸中"1"的位置就是氢原子，"2"是羧基（COOH），"3"是胺（NH_2），"4"是甲基（CH_3）。按照惯例，如果

丙氨酸中2到4号原子团按顺时针方向排列,该分子就被称为右手型或D型丙氨酸;D源自希腊语dextro(右)。如果2到4号原子团按逆时针方向排列,则是L型丙氨酸;L源自希腊语laevo(左)。一个非常奇怪的现象是,地球生命在合成自身所需的蛋白质时几乎完全使用L-氨基酸,而在合成DNA、细胞壁或其他分子时几乎完全使用D-单糖。[这里有个方便记忆的口诀是"lads",表示左(L)旋氨(A)、右(D)旋糖(S)。]然而,在实验室中合成具有手性的有机物时,通常会产生等比例的左手型和右手型分子,被称为**外消旋**混合物。

虽然具有相同的化学式,不同手性的两种分子可能具有截然不同的生物化学功能。一个例子是臭名昭著的药物沙利度胺。它的右旋形式可以治疗孕吐,但左旋形式会导致严重的胎儿畸形。另外一个更令人舒适的例子是柠檬烯香料。D型柠檬烯尝起来是柠檬味的,而L型是橘子味的。①

是什么使所有生命具有共有的手性特征,或者叫**同手性**?目前有两种流行的假说。一种假说认为,生命起源时利用的有机分子中有一种对映异构体轻微过剩,这种过剩后来被扩大。恒星的偏振光可以诱发这种对映体过剩,它可能导致有机分子在太空合成过程中产生手性过剩并传递到早期地球上;或者,放射性衰变产生的偏振辐射也可能造成有机分子的对映体过剩。这类与辐射有关的假说都存在一个问题,即手性过剩的程度通常很小,不到1%。第二种假说是,前生命化学反应只优先组合了一种对映体。例如,有机物在矿物表面的吸附可能具有

---

① 作者将此处更正为:"D型闻起来像莳萝籽或葛缕子籽,而L型是薄荷味的。"

手性选择性。相比于混合物,纯物质在化学动力学上通常更高效,所以这可能会进一步扩大上述的手性过剩。作为一个典型的生物学过程,原始的基因组复制可能要求单体具有同手性。例如,在经典的"RNA世界"假说中,复制可能需要同手性单体分子,因为只有重复的、具有相同手性的RNA链才能与RNA模板链配对。

最早的生命迹象

虽然我们对于生命起源的过程仍不清楚,但在地球上最古老的沉积岩中,我们发现了一些可能的生命活动迹象。绝非偶然的是,这些(与火山熔岩流混合的)岩石可以追溯到大约38亿年前的晚期重轰炸末期。这些岩石分布在格陵兰岛西南腹地的伊苏亚地区。它们看起来有点像《魔戒》中的魔多,阴暗深邃,给人以不祥的感觉,不过和魔多相比要寒冷得多。它们以蕴藏着诸多早期地球的信息而闻名。这些沉积岩中有被水流磨圆的砾石,还有含有石墨薄片的深海沉积物。碳同位素分析表明,这些石墨中的碳原子曾经是海洋微生物身体的一部分。碳有两种稳定的同位素,即碳-12和碳-13。生命组织中往往更富集碳-12原子,比碳-13高几个百分点,因为在生物化学上,含有更多较轻碳-12原子的分子反应速率更快。伊苏亚沉积岩中的石墨富集了大约2%的碳-12,与现代海洋微生物中发现的比例相近。由此推断,早期海洋微生物死亡后落入沉积物中,随后这些生物碳被压缩并转化为我们现在观察到的石墨。

在澳大利亚西北部,人们发现了保存更完好的早期生命证据,可追溯到大约35亿年前。20世纪80年代,我的同事罗

图2 左：澳大利亚诺思波尔德雷泽岩层中世界上最古老的叠层石化石的横截面。右：叠层石层理面的俯视图，可见叠层石顶部。参照物镜头盖的直径为5厘米

杰·别克在澳大利亚西北部的诺思波尔（North Pole）[①]附近发现了这个时期的叠层石化石（图2）。一个爱开玩笑的澳大利亚人把这个地方命名为"北极"，因为它是地球上最热、阳光最炽烈的地方之一。叠层石是一种层状沉积结构，由光合微生物群落在可接收到阳光的浅水区形成。叠层石通常是由弯曲层片构成的拱形。大片的微生物薄层，即**微生物席**，朝着阳光逐层向上生长，被包裹的沉积物最终形成了穹顶一般的叠层石。在现代叠层石上，有生命的微生物席只是一层大约一厘米厚的薄层，有着豆腐一样的质地，生长在堆积的矿物层上。如果你能旅行回到太古宙时期的地球，你会看到世界各地的海岸都遍布叠层石。今天，（在太古宙并不存在的）某些鱼和蜗牛会以这些微生物席为食，因此叠层石已经很少见了。现代叠层石只在潟湖或湖泊中分布，因为那里的水对这些贪婪的动物来说太咸了。

---

① North Pole 在英文中为"北极"之义。

现在仍然有人怀疑，在澳大利亚诺思波尔地区观察到的这些结构是否可以在没有生命的情况下形成，因此并不是真正的叠层石。但是，大多数天体生物学家都认为它们是生命遗迹，因为它们具备我们所期待的各种生物成因的特征。这些化石的叠层有着极其不规则的褶皱，很难解释为纯粹物理过程产生的结构。此外，在褶皱凸面顶部，叠层会变厚，这与光合微生物向阳光更多的方向生长时发生的情况一致。在褶皱之间的凹槽中存在一些碎片，可能是被海浪撕裂的微生物席碎片。这些碎片中的薄褶皱层含有有机碳。这些观测结果综合起来表明，澳大利亚诺思波尔地区的叠层石确实是肉眼可见的最古老的化石结构。

为了发现古老生命存在的证据，我们还可以寻找单细胞生物的个体化石，因为在地球遥远的历史时期只有微生物存在。在古老的燧石（一种细粒的、富含硅的岩石）和页岩（由泥浆形成的细粒沉积岩）中，人们发现了这种只有在显微镜下才能看到的**微体化石**。

地质学家从可能保存有微体化石的环境中采集岩石样本并带回实验室。在那里，他们把岩石样本切片，想在显微镜下碰碰运气，寻找目标。与此同时，让人头疼的如何区分生命和非生命的问题再次显现，因为有时候矿物颗粒看起来像细胞化石。因此，当在镜下观察到细胞分裂、群体行为或细胞发生连接时才会出现的丝状结构等特征时，这些微体化石才更可信。除此之外，也可以通过测试是否含有有机碳来分辨它们是不是微体化石。

目前，人们发现的最古老的、无可争议的微体化石出现在南非，时间为25.5亿年前。它们之所以令人信服，是因为它们是在细丝状的、含有有机碳的微生物席化石中被发现的，其中包括一些在细胞分裂过程中被固定保存下来的微生物形态。在南非还发现了更古老的潜在微体化石，可追溯到35亿至32亿年前。这些微体化石是球形的，含有有机碳，表现出细胞分裂的特征，但没有其他生物成因的属性。在澳大利亚西北部斯特雷利池岩层中，人们发现了34亿年前的砂岩并观察到了类似的微体化石群落，它们含有有机碳并且具有微生物活动产生的同位素特征。

世界上最具争议的早期微体化石来自澳大利亚西北部的埃佩克斯燧石岩层，距今35亿年。20世纪90年代，加州大学洛杉矶分校的比尔·朔普夫将这些微体化石鉴定为"世界上最古老的微体化石"。这些结构呈现为黑色到深棕色的条纹，似乎被分隔成类似细胞的结构。朔普夫给它们起了拉丁文名字，暗指它们是一种已经灭绝的蓝细菌（一种光合细菌）。2002年，牛津大学的马丁·布拉西耶重新检查了这些样本，发现它们在三维空间上并不像微生物细丝。这引起了一场激烈的争论。布拉西耶还指出，这些燧石不是浅层沉积形成的，而是在热液喷口产生的，这是一个不太可能找到需要阳光的蓝细菌的地方。后来，进一步的研究表明，这些细丝实际上是被赤铁矿（一种氧化铁）部分填充的裂缝，所以呈深色。然而，人们也在细丝外的燧石中发现了有机碳，因此关于这些样本是不是微体化石的争议仍在继续。

到目前为止，我们已经讨论了可以利用同位素、微体化石和叠层石来寻找最早的生命痕迹。除了这些之外，我们也可以利用**生物标志物**来指示早期生命的痕迹。生物标志物是生物分子的可识别衍生物。我们都知道，化石骨架让我们能够区分霸王龙和三角龙。在微观尺度上，微生物可以留下独特的有机残留物，往往由环状或链状有机碳组成。这些特殊的"分子骨架"来自只在特定类型的微生物中发现的特定分子。因此，生物标志物不仅可以证实过去生命的存在，还可以指示特定的生命类型。

遗憾的是，就像上面提到的一些微体化石一样，对于古代生物标志物的解读也陷入了争议。已报道的最古老的生物标志物大约可以追溯到28亿年前，并且似乎存在产氧蓝细菌产生的有机分子信号。但是，后续的研究发现，这些采集的岩石很可能被取样过程中带入的更年轻的有机物质污染了。相较而言，比较确定的早期生物标志物出现在一些25亿年前的岩石中。这些生物标志物位于看起来未受污染的微小流体包裹体中。

尽管一些证据有局限性，但地质记录总体显示地球在35亿年前已经出现了生命。此时，地球刚形成还不过10亿年，也就是在大撞击事件发生后不久。这表明，在地质时间尺度上，生命也许很快就可以在宜居行星上诞生。这意味着，即便我们认为火星上宜居环境存在的时间比较短（详见第六章），那里也不是不可能曾经演化出生命。

第四章

从史莱姆[①]到史莱克

地球如何维持适宜生命生存的环境？

　　无论生命是如何形成的，它从诞生起已经存在了超过35亿年，并且从简单的微生物软泥逐渐演化到了如今复杂的人类文明。在这段漫长的时期，尽管太阳光的量增加了约25%，但地球始终维持着海洋并且在大部分时间里保持着适宜的气候。太阳逐渐变亮是处于主序阶段的恒星的演化结果。当太阳核心的四个氢原子核发生聚变形成一个氦原子核时，总的粒子数会减少，所以太阳核心外围的物质会向内压缩以填补空余的空间。核心的压缩会伴随温度升高，导致聚变反应进行得更快。因此，太阳的亮度每10亿年就会增强7%到9%。目前，科学家通过对处于不同演化阶段的类太阳恒星的观察，证实了这个理论。

　　如果20亿年前或者更早期的地球拥有和现在类似的大气

[①] 史莱姆英文为slime，有"软泥"之义。

成分，那么当时比较暗淡的太阳辐射会导致整个地球处于冰封状态。但是，地质证据表明事实并非如此。这个谜题也被称为**暗淡太阳悖论**。目前有一些证据显示，至少在38亿年前早期地球表面就存在稳定的液态水。比如，当时已经出现各种沉积岩，而沉积岩的形成依赖液态水将物质从陆地搬运到海洋中。

关于暗淡太阳悖论，目前有三种解释。其中最有可能的一种解释是，早期地球存在更显著的温室效应。另外一种解释认为，早期地球整体比现在更暗，因此吸收了更多的阳光，然而这个说法缺乏有力的证据。第三种解释认为，早期太阳会以快速外流（太阳风）的方式向太空释放大量物质，因此太阳的核心就不会像上文假设的那样随着时间的推移被压缩和加热。如果太阳外流的质量合适，那么早期太阳的亮度可能和今天是一样的。然而，目前对于类似早期太阳的年轻恒星的观测结果并不支持第三种假设。

在评估第一种假设时，我们需要认识到大气使地球变暖的程度取决于大气成分。如果没有大气（并假设地球反射的太阳光的量保持不变）的话，地球表面将低至−18℃。而现今全球表面平均温度是+15℃。这33℃（=18℃+15℃）的温差便缘于地球的**温室效应**，即由大气导致的变暖。

那么，温室效应是如何产生的呢？当行星表面受到可见光照射而升温时，它会发出红外线，就像我们温暖的身体向外发射红外光。大气对来自其下部的红外辐射具有显著的阻挡作用，并吸收了这些热辐射的能量而升温。同样地，升温后的大气也会发出红外辐射，而且这些来自大气的红外辐射中的一部分会回到地表。因此，有大气的行星表面比没有大气的行星表面要更温暖，因为它会收到可见光以外的、来自大气辐射的额外能量。

在35亿年之前，太阳亮度比现在要低25%。而现在地球大气温室效应提供的增温为33℃。因此，要使地球早期平均温度维持在接近现在的温度水平，需要大气温室效应贡献约50℃的增温。当大气中含有浓度更高的温室气体时，它们会吸收更多地表红外辐射，因此导致更强的温室效应。当前，地球大气产生的33℃温室效应中，大约三分之二是由水蒸气造成的，其余三分之一中的大部分则是由二氧化碳造成的。然而，水蒸气会凝结成雨或雪，其浓度基本上取决于受大气中二氧化碳浓度变化控制的背景温度，因为二氧化碳不会凝结，始终是气体。因此，尽管二氧化碳在地球大气中的含量较低，但它依然控制着当前温室效应的强弱。1700年前后，地球大气中二氧化碳的水平约为280 ppm（即大气的每100万个分子中，有280个是二氧化碳分子）。然而，在2010年，二氧化碳含量升高到390 ppm。这主要是因为自工业革命以来，森林砍伐和石油、煤炭等化石燃料的燃烧释放了大量的二氧化碳。科学家通过对**古土壤**（也就是化石化的土壤）的化学分析，推算出太古宙末期二氧化碳含量最高可比工业革命前水平高出数十倍至一百倍。但即使是这么高的二氧化碳水平产生的温室效应，也不足以抵消早期较弱的太阳亮度导致的低温。

事实上，太古宙大气中氧气的含量不到1 ppm，导致甲烷可能成为重要的温室气体。现代大气中第二丰富的气体是氧气，占21%（第一丰富的氮气占78%，但它和氧气一样都不是温室气体）。它会与甲烷反应，导致现代大气中甲烷含量很低（1.8 ppm）。但是在太古宙，无氧的大气导致甲烷可以积累到数千ppm的水平。

然而，甲烷也并不会无限制地积累，因为它会在高层大气中被紫外线分解。甲烷的光化学分解产物可以继续反应生成其他碳氢化合物（即由碳和氢组成的化学物质），包括乙烷（C_2H_6）气体以及含有煤烟颗粒的烟雾。甲烷、乙烷和二氧化碳的混合物，加上由这些温室气体导致的升温所产生的水蒸气，可以产生足够强的温室效应来抵消暗淡太阳造成的影响。这个理论假定了太古宙时期微生物可以像现在这样源源不断地提供甲烷，而这确实也是可能的，因为现在普遍认为地球很早就出现了生成甲烷的微生物代谢活动（详见第五章）。

当然，我们可能会问在生命出现之前地球上的气候是什么样的。在那种情况下，二氧化碳可能主导着地球大气的温室效应，就像现在一样。事实上，在地球历史上的大部分时间中，二氧化碳的地质循环在大约百万年的时间尺度上调节着气候。（这种调节机制现在也在发挥作用，只是相比于人类导致的全球变暖来说慢了许多，不足以在短期内抵消人类活动的影响。）这种二氧化碳的地质循环大体如下。大气中的二氧化碳溶解在雨水中，并与大陆上的硅酸盐岩发生化学反应。然后，**化学风化**释放的溶解碳随着河流被输入海洋，并在那里沉淀形成石灰岩（$CaCO_3$）等含碳沉积物。如果大气中的二氧化碳最终都沉积为碳酸盐岩，那么地球大气将失去温室效应，地球将最终冰封。然而，地球也有一种过程，可以将沉积的碳酸盐岩中封存的碳重新释放到大气中。海底的碳酸盐岩可以被缓慢移动的大洋板块搬运到板块边界，随后**俯冲**到其他板块之下。现在发生的一个例子是，南太平洋的纳斯卡板块向东俯冲到智利下面。在板块俯冲过程中，碳酸盐岩受到挤压并被加热，从而分解成二氧化碳。这部分二氧化碳可以由火山作用（岩石熔化）和变质作用（岩石受热承压但不熔化）重新释放进入大气。大气二氧化碳的损失和补充过程构成的整个循环被称为**碳酸盐-硅酸盐循环**。这个循环就像一个地球表面温度的自动调节器。如果气候变暖，更多的降雨和更快的风化会消耗更多的大气二氧化碳，使地球降温。如果地球变冷，空气会变得更干燥，二氧化碳通过风化而固定下来的速度就会变慢，如此，地质过程中释放的二氧化碳会逐渐在大气中积累起来，从而增强温室效应，使地球变暖。

　　这种碳酸盐-硅酸盐循环可能早在生命起源之前就已经开始调节地球气候了。在整个地质历史上，地球大气发生过两次主要的氧气剧烈升高事件，分别在约24亿年前和7.5亿至5.8亿年前。大气中氧气的含量升高，甲烷的浓度随之降低，使得碳酸盐-硅酸盐循环在调节地球气候方面发挥了更加重要的作用。值得注意的是，上述循环在冥古宙和太古宙时期可能是不同的。这主要是由于**板块构造**——地质板块在地幔内部对流单体之上的大规模运动——可能具有不同的模式。地幔中的放射性元素在衰变时会产生热量，而早期地球有更多的放射性元素。这可能会通过下面两个方面影响地幔对流方式。一方面，更热和更软的冥古宙地幔使得海洋地壳可以更快地俯冲下沉。另一方面，冥古宙更热的地幔更易熔融，导致海洋地壳更厚且温度更高，不利于俯冲。总之，冥古宙时期花岗岩的存在（从锆石推断出的，详见第三章）意味着原始地壳一定是以某种方式被埋藏了，因为花岗岩是在下沉地壳熔融时产生的。然而，早期地球的地壳活动仍然是一个悬而未决的问题。

大氧化事件:迈向复杂生命的一步

地球大气成分经历过的最剧烈变化是氧气的增加。氧气和温室气体的变化对地球生命的演化均发挥着至关重要的作用。在地质历史的大部分时间里,大气中的氧气含量都很低,无法支持呼吸氧气的动物出现。24亿至23亿年前的**大氧化事件**第一次氧化了地球大气。然而,大氧化事件后大气中氧气含量仅达到0.2%至2%体积占比,远低于今天21%的水平。此后,一直到约5.8亿年前,第二次增氧使大气中氧气超过3%后,地球上才出现了大型动物(图3)。

现今地球大气中几乎所有氧气都是生物产生的。极少部分的氧气可以在不需要生命参与的情况下产生。例如,紫外线照

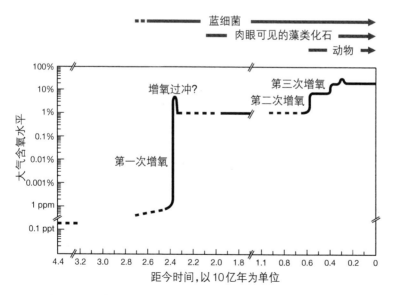

图3 基于地质证据推断的大气氧气的大致历史(ppm:百万分之一;ppt:万亿分之一)

射会使上层大气中的水蒸气分子发生光化学分解,形成氢和氧。光化学分解产生的氢原子可以快速地逃逸到外太空,而氧原子则被留下来,从而阻止了氢和氧重新结合生成水。但这种方式产生氧气的量很小,主要是由于上层大气十分干燥。相比之下,生物活动可以通过**产氧光合作用**产生大量氧气,成为大气中氧气的主要来源。植物、藻类和蓝细菌可以利用太阳光的能量将水分解成氢和氧,随后将氢和二氧化碳结合生成有机物,而氧气则被作为废物释放。

另一种更原始的微生物光合作用类型是**不产氧光合作用**,即它不能分解水和释放氧气。在这种光合作用中,原始微生物会利用太阳光和氢气、硫化氢或火山周围热液区溶解的铁来合成自身所需的有机物。现今,大家在温泉中看到的微生物浮沫就是这样生长的。

在植物和藻类出现之前,最早的产氧生物与现代的蓝细菌类似。现今的海洋和湖泊中充满了这种蓝绿色的细菌。DNA研究表明,一种类似于蓝细菌的微生物是植物、藻类和现代蓝细菌的共同祖先。因此,我们可能会以为,一旦地球演化出了蓝细菌,大气将会被迅速氧化。然而有证据表明,在大气被氧化之前的很长一段时间里,早期的蓝细菌就已经开始产生氧气。那么为什么大氧化事件的发生会滞后于产氧光合作用?一种合理的解释是,地表的**还原性物质**可以快速地消耗产氧光合作用释放的氧气。这些还原性物质包括来自火山、地热地区和海底热液喷口的气体,如氢气、一氧化碳和甲烷。此外,早期海水中还有含量较高的还原态的铁离子,也可以与氧气反应。太古宙独特的富铁沉积岩——**条带状铁建造**——表明,太古宙海洋中有大量的溶解态铁,与现代海洋大相径庭。

生物光合作用产氧的最早一批迹象出现在大约28亿至27亿年前,主要证据来自叠层石化石和一些与氧气反应会变得更易溶的化学物质。在澳大利亚西北部,坦比亚纳组岩石中的叠层石曾环绕着古湖泊生长。理论上,进行不产氧光合作用的微生物也可以形成这样的叠层石结构,但是目前并没有证据显示有热液活动可以为这些不产氧光合作用提供关键的反应底物。更有可能的解释是,进行产氧光合作用的蓝细菌可能形成了这些叠层石结构,这也符合从叠层石中观察到的疑似因为蓝细菌丝朝向阳光生长而产生的从簇与尖顶等结构特征。除此之外,一些在被氧化后才溶解释放的元素,如钼和硫,自28亿年前开始在海洋沉积物中富集,这可能是由微生物在陆地上利用如叠层石产生的局部氧气源氧化这些元素造成的。

氧气没有立即在地球大气中积累的另一个原因是,它的产生基本上是一个可逆的过程,即当一个氧分子生成时,一个有机碳分子也会按照下面的反应式随之生成:

$$二氧化碳+水+阳光=有机物+氧气$$
$$CO_2+H_2O+阳光=CH_2O+O_2$$

这一过程很容易逆转,即有机物也可以与氧气发生反应,再生成二氧化碳和水。此外,太古宙大气无氧意味着微生物可以很容易地将有机碳转化为甲烷气体,就像今天在臭臭的、无氧的湖泊或海底沉积物中发生的那样。在空气中,甲烷可以与氧气发生反应,生成水蒸气和二氧化碳。无论是上述哪种情况,产氧光合作用的发生都没有办法使得氧气净增加。

然而，有一小部分有机碳（对于当前的地球来说，约为总碳的0.2%）会被埋在沉积物中，从而与氧气隔离开，导致这部分有机物未被氧化。这样一来，每埋藏一个有机碳原子都等于释放出了一个氧分子。当然，这种"释放"的氧很容易与除有机碳以外的许多物质发生反应，包括地质活动产生的多种还原性气体和可溶性矿物质，比如铁。尽管如此，当进入大气的还原剂的通量低于通过有机碳埋藏净释放的氧气时，大气的氧化还原平衡就会达到一个临界点，并导致大气的氧化。在大氧化事件之后，早期大气中氧气含量在迅速达到0.2%至2%的水平后却并没有持续增长。这可能是因为溶解在雨水中的氧气基本上都与陆地矿物反应而消耗掉了，从而阻止了氧气进一步积累。

大氧化事件带来的一个显著变化是氧化反应激增。例如，大氧化事件之后，"红层"，即铁矿物与氧气反应生成的铁锈，开始普遍出现。这种"红层"现在也很常见，比如在美国西南部。大气的氧化也导致大气中硫的转化反应发生了转变。在大氧化事件之前，大气中的氧气含量还不到百万分之一（体积占比）。在这种无氧的大气中，火山喷出的二氧化硫在几千米的高空被强紫外线分解后，会以红色或黄色的单质硫颗粒形式沉降到地表。这些单质硫颗粒记录了光化学反应产生的特殊同位素特征[①]，并且在无氧条件下可以稳定保存在岩石中，指示了大气的组成。然而，大氧化事件之后，大气中的游离氧可以直接与硫结合，阻止了单质硫颗粒的形成。自此，天空不再有硫沉降下来，同时沉积岩中硫的特殊同位素特征也消失了。

---

① 此处指硫同位素的非质量相关分馏特征。

伴随着大气的氧化，地球平流层中的臭氧层也形成了。现今，地球上的臭氧聚集在距离地表20至30千米之间的区域，保护着地球表面的生物免受紫外线的伤害。如果没有臭氧层，我们会在几十秒内被严重晒伤。臭氧是由三个氧原子组成的分子，这三个氧原子均来自氧气。具体的形成过程是，大气中的氧分子（O_2）被阳光分解成两个氧原子（O），然后氧原子分别与其他氧分子结合形成臭氧（O_3）。

到目前为止，有关大氧化事件的成因仍然众说纷纭。然而无论是哪种观点，都必须归结为氧气产量因有机碳埋藏而增加或氧气消耗量减少。第一种观点似乎与现有的观察不符合。当生物合成有机碳时，它们偏向于从海水中摄取轻碳，即碳-12，因此有机碳埋藏的增加往往伴随着海水中轻碳占比的降低。但古代灰岩中记录的海水碳同位素组成并没有显示出碳-12占比持续降低。沉积记录似乎表明并没有大量的有机碳埋藏。在这种情况下，第二种假说变得更有可能。

然而，是什么导致了耗氧物质输入通量减少呢？一种可能性是，早期大气中含量较高的富氢气体，例如甲烷和氢气（在没有氧气的大气中很常见），会导致氢原子迅速逃逸出大气层。大气中氢的持续逸失逐渐氧化了固体地球，地球岩石中还原性物质占比降低，导致产生更少的还原剂来还原生物产生的氧气。在这个过程中，无论是甲烷还是氢气，大气上层逃逸的氢原子最终来源都可以追溯到水分子。从理论上来说，当氢原子逃逸到外太空后，其来源处的氧原子就剩余下来。在这个过程中，水蒸气本身很难到达上层大气，因为它会随着温度降低凝结成云。但是，其他富氢的气体，例如甲烷，并不会凝结，可以把氢原子带到上层大气并在那里逃逸。

我们可以用高中时将酸逐渐滴入碱性溶液的酸碱滴定法来粗略类比大氧化事件。在滴定过程中,溶液的颜色会在某一刻突然发生变化,通常是从无色变成红色。古代大气也经历了类似的转变,只不过不是颜色的变化,而是从富氢变成了富氧。这里,我们可以把地球古代大气的这种"滴定"称为**"氧化还原滴定"**,它是**还原剂**(例如氢气)和**氧化剂**(例如氧气)之间的竞争。

"枯燥的十亿年"的结束开启了动物时代

大氧化事件之后,海洋和陆地逐渐适应了大气氧化带来的变化。在大约18亿年前,大气中的氧气浓度逐渐稳定,并在很长一段时间内保持在0.2%到2%之间(体积占比)。与此同时,复杂生命的演化速度也比较缓慢。这或许是由于当时的海水整体呈现为表层中度氧化但表层以下无氧且富含铁或硫化氢。对于复杂生命来说,这种无氧的水环境是有害的。18亿至8亿年前,复杂生物的演化出奇地缓慢,以至于有人将这段时间称作**枯燥的十亿年**。就像一篇科学论文中写的那样:"在如此长的时间里如此缺乏生物演化事件,这在地球的历史上从来没有发生过!"

实际上,在这所谓的"枯燥的十亿年"前后,发生了一些重要的生物演化事件。大约在19亿年前,第一批肉眼可见的生物单体化石出现了。这些直径几厘米、具有螺旋纹的化石可能是海藻,被称为卷曲藻。在中国,从18亿年前的沉积岩中发现了疑源类化石。这是一种大于0.05毫米且具有有机壁的化石。有些疑源类化石被认为是藻囊,即藻类在干燥时期变成不活动的球时形成的结构。此外,这10亿年中也出现了新的生物谱系,包括从12亿年前的地层中发现的红藻化石,代表生物可能演化出了有性繁殖。如今,我们把古老红藻的后裔煮过后做成琼脂,用来让冰淇淋变稠。我们还用一种被称为"海苔"的红藻来做寿司。总之,有性繁殖的演化让这"枯燥的十亿年"不至于**那么枯燥**。

在元古宙末期,地球发生了第二次大氧化事件,从约7.5亿年前开始,到约5.8亿年前达到顶峰。其间,大气氧气含量发生了第二次升高,达到超过3%(体积占比)的水平,同时,深海也完全被氧化了。在距今约6.3亿年之前的地层中,出现了动物活动产生的生物标志物和(可能是海绵的)微体化石。但直到在距今5.8亿年之后的地层中,我们才观察到厘米到米级的宏体生物化石。这是一些奇怪的软体生物,没有嘴巴或肌肉,必须依赖扩散穿过表皮的营养物质而生存。它们有些看起来像披萨,而有些看起来像长条状的植物叶子。然而,这些生物生活在海洋深处,无法接受阳光,所以它们不可能是植物。它们被统称为**埃迪卡拉生物群**,因为它们最初被发现于澳大利亚南部的埃迪卡拉山。埃迪卡拉生物群持续了数千万年后灭绝。

埃迪卡拉纪之后紧接着发生的是**寒武纪生命大爆发**。在距今5.41亿年[①]的寒武纪之初,许多具有新型身体结构的动物化石在短短1 000万至3 000万年间骤然出现。大多数现代动物的祖先开始出现,其中包括许多具有坚硬骨骼的动物。寒武纪(距今5.41亿至4.88亿年)[②]出现在元古宙结束后,是显生宙

---

① 根据最新定年结果,应修改为"距今5.39亿年"。
② 根据最新定年结果,应修改为"距今5.39亿至4.85亿年"。

（距今5.41亿年前至今[①]）的第一个阶段。显生的意思是"可见的动物"，源自希腊语phaneros（可见的）和zoion（动物）。寒武纪时期出现了一种小生物皮卡虫，它是一种只有几厘米长、可以游泳的蠕虫。皮卡虫所在的生物分支可能最终演化出了包括我们在内的所有脊椎动物。

  由此可见，大气和海洋的第二次氧化极大地增加了动物多样性，然而导致这次氧化事件的原因仍然没有定论。有的观点认为，多种因素导致了此时期有机碳埋藏的增加（从而增加了氧气的产量）。例如，陆地上新物种的出现促进了表面岩石的分解和养分的释放、海洋中演化出浮游动物，以及黏土矿物的大量形成帮助埋藏有机物。另一个可能的原因是氢原子持续逃逸。在"枯燥的十亿年"，大气中可能含有中等浓度（大约100 ppm）的甲烷，导致更多的氢原子逃逸到太空，从而促进了海底的氧化，并且在第二次全球"氧化还原滴定"之后，提高了大气中氧气的浓度。

  在大约4亿年前，维管植物开始登陆，并发生了第三次大气氧浓度的升高。这些植物可以利用它们特殊的组织（如木质素）来运输水和矿物质。维管植物通过根劈作用促进了地表岩石中养分的释放与有机物的埋藏。此后，大气中的氧气水平一直保持在10%到30%（体积占比）之间，有利于陆地动物的延续，最终我们人类出现。

"雪球地球"还是"水带地球"？

  一个比较有意思的现象是，第一次和第二次大氧化事件都

[①] 根据最新定年结果，应修改为"距今5.39亿年前至今"。

伴随全球范围的冰川活动，就好像"枯燥的十亿年"两端的书立。一些科学家认为，海洋在这些时期处于整体冰封的状态，地球被称为**雪球地球**。这种地球状态远比近几百万年间经历的更新世冰川期更极端。在更新世冰川期，冰川活动只是延伸到了中纬度地区，并没有到达热带地区。为了叙述方便，我们在这里讲解一下地质时期划分。元古宙划分为三个代：古元古代（距今25亿至16亿年）、中元古代（距今16亿至10亿年）和新元古代（距今10亿至5.41亿年）。地质学家往往把古元古代早期（距今24.5亿至22.2亿年）发生的三四次冰期事件统称为"古元古代冰川作用"。而"新元古代冰川作用"是指两次大冰期事件（距今7.15亿至6.35亿年）和一次较小的冰期事件（距今5.82亿年）。

这些全球范围的冰川活动留下了一些岩石记录。例如，冰盖的流动会使石头、沙子和泥巴混合形成一种被称为**冰碛岩**的特殊岩石。冰川在流动的过程中还会在其下方的岩石上留下平行的划痕，被称为冰川擦痕。此外，一些岩石会从冰山或融化的冰架上掉落，砸在沉积物上导致其变形，这些岩石在其后形成的沉积岩中表现为**坠石**。一些沉积记录中发现的冰碛岩、**冰川擦痕**和坠石都是冰川曾经存在的确凿证据。

20世纪80年代和90年代，科学家们对岩石记录的古磁场进行了测试分析。结果表明，冰川活动曾经延伸到了热带地区。我们都知道，指南针可以指示地球南北方向的磁场。一个鲜少有人知道的事实是，磁力线近似垂直于两极附近的地表，而平行于赤道附近的地表。火山岩中含有一些富铁矿物，如磁铁矿（一种氧化铁，Fe_3O_4）。当火山岩浆冷却时，其中的磁铁矿会沿着地球磁场的方向被磁化。因此，当火山岩中磁铁矿捕获或剩余的磁场线与古岩层平行时,我们就知道这些岩石形成于热带地区。这种现象出现在新元古代和古元古代冰期形成的带有冰川活动痕迹的岩石中,表明当时存在全球性的冰川活动。

如果海冰可以蔓延到热带地区,那么整个地球应该都处于冰封的状态。这个想法起源于20世纪60年代。当时,俄罗斯气候学家米哈伊尔·布德科通过计算发现,如果极地冰盖向赤道延伸超过南北纬30°,整个地球都会被冰封。冰可以反射大量的阳光,因此具有很高的**反照率**,即照射到地球上的阳光被反射的比例(在0到1之间)。如今,地球的反照率约为0.3,这意味着30%的阳光会被反射到太空中。但是,海冰本身可以反射50%的入射阳光,而覆盖了雪的海冰可以反射的则高达70%。当热带地区出现高反照率的冰时,地球吸收的阳光会更少,因此温度更低,进而会形成更多的冰并导致温度进一步降低,最终表现为灾难性的**冰反照率失控现象**。总的来说,一个被冰封的地球将表现为:平均气温在−35℃以下,赤道地区的海洋被厚达1.5千米的冰层覆盖,两极被3千米的冰层覆盖。

加州理工学院的乔·克里什芬克提出了一个理论来解释"雪球地球"是如何最终融化的。他也是首次提出**雪球地球**这个名词的科学家。他认为,火山活动可以打破冰层,将二氧化碳排放到大气中。因为"雪球地球"时期没有降雨,也没有广阔的海洋来溶解二氧化碳,所以大气中的二氧化碳会逐渐积累。经过几百万年的积累,高浓度的二氧化碳会产生强力的温室效应,最终将"雪球"融化。此外,冰川在从极地较厚冰层向热带较薄冰层生长的过程中,将携带火山灰和被风吹起的灰尘,导致热带地区的海冰颜色较暗,可以吸收更多的太阳光,使"雪球"更容易融化。在这种情况下，冰反照率失控现象就会发生逆转：温室效应使得全球变暖，冰川融化加快，全球冰量减少会使地球进一步变暖，从而导致更多的冰川融化，如此循环直到达到下一个稳定状态。

人们在冰川沉积物之上发现了厚厚的碳酸盐岩沉淀，它被称为**盖帽碳酸岩**，似乎支持了上述"雪球"融化机制。人们猜测，随着覆盖全球的冰融化，大气中积累的大量二氧化碳最终被转化为这些盖帽碳酸岩。"雪球地球"假说的支持者们还观察到，盖帽碳酸盐中碳-12和碳-13的同位素比值与火山喷发产生的二氧化碳中的比值相似，进一步支持了这一假说。

然而，"雪球地球"假说的一个核心问题是：光合作用藻类的代谢依赖阳光，它们如何在厚厚的冰层下还能幸存。而绿藻和红藻的谱系在新元古代的"雪球地球"时期并没有中断。一种假说认为，火山活动释放的热量可以导致其周边的海冰融化或变薄，形成透光的局部区域，从而维系藻类的生存。但也有人质疑，在"雪球地球"时期，整个地球是否真的都被冰层覆盖了。例如，一些科学家利用相对复杂的计算机模型，模拟了地球早期气候特征。结果表明，即使是在冰期，低纬度地区仍可能存在一片带状的开放水域。这个发现也被称为**"水带地球"假说**。"水带地球"的存在主要是因为低纬度地区会被颜色较暗的裸冰覆盖，其反照率相对较低，而反照率高的雪冰只存在于高纬度地区。

无论是上述哪种冰期状态，新元古代和古元古代冰川作用产生的原因可能都是地球上温室效应的减弱。在"雪球地球"时期之前，板块运动导致的大陆漂移和重组使得陆地在赤道地区聚集成了一片超大陆。之后，赤道地区的频繁降雨可能会增强大陆风化，从而将大气中的二氧化碳降低到正常水平之下，有利于地球进入冰期。此外，人们也观察到"雪球地球"与氧气含量的升高同时发生，表明大气中的另一种温室气体——甲烷——含量的变化可能是另外一个重要诱因。大气中氧气和甲烷可以发生反应，因此它们无法同时达到较高的浓度。在"雪球地球"时期，地球大气中氧气含量的升高可能导致了甲烷含量的快速降低，从而减弱了温室效应。

对于天体生物学研究来说，"雪球地球"的启示意义在于，类地行星上形成的生物圈是有一定韧性的。即使面临巨大的气候变化，地球上的生命仍然在新元古代大冰期事件后幸存下来，并且演化进入了一个新的动物时代。

复杂生命的出现

在天体生物学研究中，我们经常好奇，动物类型的生命是否存在于银河系的其他地方。这种可能性或许可以从地球上生命的演化历史中得到启示。地球上的生物若想从简单变得复杂，必须克服至少两个关键障碍。首先，生命必须演化出复杂的细胞，从而在更大尺度上实现细胞的分化，例如产生肝脏细胞和脑细胞。大型的、具有三维结构的多细胞动物和植物只能由**真核**细胞构成，这是因为真核细胞可以发展出更多样的细胞类型（详见第五章）。

产生复杂生物的第二个前提条件是，大气中含有足够的氧气，使得生物可以像我们人类一样通过需氧的**有氧**代谢产生大量能量。生物长大和移动需要大量的能量来支撑。在摄入相同量的食物时，不需氧的**厌氧**代谢产生的能量只是有氧代谢的十分之一。然而，对于地外生命来说，我们可能会想象，它们是否可以利用氟气或者氯气这两种强氧化剂来作为氧气的替代品，从而进行高能量产出的代谢。答案是否定的。氟气的反应活性很强，与有机物在一起会爆炸；而氯气溶于水后会生成对生物有害的漂白剂。对于复杂生命来说，氧似乎是元素周期表中的最佳选手，它具有足够高的反应活性，但又不至于太过强烈。

然而，多细胞生物的出现不仅依赖氧气的存在，而且需要氧气含量达到一定水平。最初的好氧多细胞生物可能是细胞在分裂过后无法分离从而产生的细胞聚集体。这些细胞聚集体的体积会受限于外层氧气向内层细胞的扩散，从而无法达到较大的规模。因此，更高的大气氧气含量可以支持更大规模的细胞聚集体。假设细胞都具有典型的代谢速率，这种受氧气扩散限制的原始动物需要超过3%体积占比的大气氧气含量才能够维持几毫米的大小。这正好对应了约5.8亿年前发生的第二次大气氧化事件后氧气的水平，为后生动物的出现奠定了基础。

同时，藻类化石记录表明，真核生物出现的时间远早于动物。所以，寒武纪生命大爆发之前的漫长时期可能是在为动物大规模出现积累足够的氧气。我们甚至可以将大气达到复杂动物生命生活所需的氧气水平的时间定义为**氧化时间**。对于地球来说，这个时间是它形成后约40亿年。

相比而言，类地系外行星的氧化时间仍然是不确定的，因为我们仍然在试图解答是什么决定了地球的氧化时间尺度。归根结底，大气中的氧气来自液态水。因此，具有海洋的类地系外行星在演化出产氧光合作用之后，有可能最终形成富氧的大气。如果某个系外行星含有很多可以与氧发生反应的物质，使得其大气的氧化时间超过120亿年——这将会超过类太阳恒星演化成"红巨星"的时长——那么这种类地行星很可能无法演化出复杂生物，至多只会演化出一些简单的微生物软泥。反之，大气氧化时间比较短的系外行星，可能比地球更快地演化出复杂生物。

生物演化是否有趋势？来自大灭绝的认识

我们可能会好奇，生物一旦开始演化，是否有向更复杂方向发展的趋势。以陆地上的生物为例。在海洋生物逐渐登陆的过程中，第一个植物孢子出现在距今4.7亿年左右，而大部分植物残体化石出现在距今4.25亿年左右。昆虫在距今4亿年左右开始在陆地上出现。然后，鱼类在距今3.65亿年左右演化出两栖动物；它们的后代最终演化出爬行动物和哺乳动物，包括我们人类。这里展现出一个趋势：生物圈逐渐学会更大程度地利用地球资源。但是，把演化想象成生物逐步演化出我们现在的生物圈是一种误解。事实上，化石记录表明，有许多物种出现，也有很多物种消失。此外，一些生物灭绝事件也会间歇性地减少复杂生命的多样性。例如，地球历史上发生过多次大灭绝事件，导致超过25%的科（科是包括属和种的更高级生物分类单位，详见第五章）消失。当然，这些大灭绝事件指的是生物圈中非微生物部分的灭绝。

生物的多样性在元古宙-显生宙的过渡时期才显著增加。因为在那个时期，生物获得了新的身体结构和生存方式，并得以保存至今。对于动物来说，一个关键的身体结构革新是出现了**体腔**。体腔中可以充满液体，从而增强身体的刚性且利于生物集中肌肉力量。这种革新促进了动物产生自我推动能力，并且这种能力在生物先外后内演化出坚硬的骨骼后得到进一步提高。在这个过程中,一种捕食者-猎物的"军备竞赛"关系可能促进了寒武纪动物种类的大爆发。稍后,植物的多样性也随着维管系统(包括负责运输的刚性细胞以及其他相关组织)的出现而增加,并最终演化出了树木。

在过去的5亿年里,地球共经历了五次生物大灭绝事件,导致超过50%的物种消失。其中,两次最大的灭绝事件分别发生在距今2.51亿年的二叠纪末期和距今0.65亿年的白垩纪末期。二叠纪末期大灭绝事件是二叠纪(距今2.99亿至2.51亿年)和三叠纪(距今2.51亿至2亿年)的划分标志。这一事件杀死了除鳄目动物以外所有比家犬体型更大的陆地动物,给海洋生物也带来了巨大的灾难。例如,这次事件彻底毁灭了寒武纪海洋中的标志性生物——三叶虫,虽然三叶虫在此事件前已经开始减少。仅次于二叠纪末期大灭绝事件的白垩纪末期大灭绝事件是白垩纪(距今1.45亿至0.65亿年)和古近纪(距今0.65亿至0.23亿年)的划分标志(一些文献按照旧的命名体系,将这一事件称为白垩纪-第三纪灭绝事件)。此次事件导致了恐龙的灭绝。二叠纪末期大灭绝事件使95%的海洋物种和70%到80%的陆地动物灭绝,而白垩纪末期大灭绝事件使地球上65%到75%的物种灭绝。

二叠纪末期大灭绝事件貌似是由地球自身活动造成的一系列不幸事件。事件的导火索似乎是西伯利亚地区发生的大规模火山活动,所涉地区面积与欧洲大陆相近。这一地区下面有煤层,火山活动引起大规模的煤层燃烧,释放了大量的甲烷和二氧化碳等温室气体,使海洋变暖并酸化。氧气在海水中的溶解度随着温度升高而降低,因此海洋变暖造成了深海缺氧。缺氧的深海可能会产生并向海面释放有毒的硫化氢气体。气候变暖、海洋酸化和有毒气体释放三者共同作用,导致此次事件成为显生宙最严重的灭绝事件。

白垩纪末期大灭绝事件的诱因可能是一颗直径约10千米的小行星撞击了地球。这次撞击带来了一系列的灾难性后果,包括暂时性的臭氧层破坏、全球范围的野火,以及酸雨。随后,更多硫酸盐注入地球平流层,这会反射更多的入射太阳光,从而导致全球气候变冷。这次撞击事件产生的撞击坑,在墨西哥小镇希克苏鲁伯地下约1千米处被发现。

这种大灭绝事件破坏了之前占据优势地位的生物谱系,但也为其他生物谱系的繁盛和演化提供了机会。直白地说,正是因为希克苏鲁伯撞击事件灭绝了恐龙,哺乳动物才能在生物圈中占据统治地位,并最终演化出人类,而你才可能读到这本书。

不过,大灭绝事件带来的另外一个启示是,一个文明可以得到发展,也可能被彻底毁灭。例如,平均每1亿年左右,地球上会发生一次可以导致白垩纪末期大灭绝这种规模灾难的撞击事件。这意味着,宇宙中随机灾难事件的发生可能会导致系外行星文明相比于宇宙寿命来说比较短暂。因此,即使类地行星一直适合生命存在,文明也可能是短暂的。这无疑会对寻找外星智慧生物产生不利影响(详见第七章)。

第五章

生命：基因组的延续和适应之道

地球的生命：从更高的角度看

我曾在第一章中给出生命的一般定义。但是更精细地审视我们现有的唯一一种生命，将有助于我们寻找其他地方的生命。鉴于此，让我们先从全球尺度来总结地球上所有生物的特征，然后再更精细地探讨细胞和分子。

现在，让我们想象一名星际旅行者来到了地球，意图了解我们的生物。也许当她在自己的星球上时，她就已经从地球湿润且异常富氧的大气层特征推断出地球上有生命。抑或是，她收听到了几十年前的地球电视广播，并且不知为何这些信号没有让她放弃来访。令人惊奇的是，就像电影中的外星人一样，她可以说一口流利的英语，而且她的飞船也碰巧在一个英语区国家着陆！我们会跟她说些什么？

在全球层面上，地球的**生物圈**是所有活生物体和死生物体的总和。有时这个术语也包含了生命所占据的非生命区域。我们可以通过数十亿吨的生物碳来量化生物圈各部分，从而确定其大致的组成成分。陆地上的生物量大约有2万亿吨碳，其中30%到50%是活的生物碳，其余是死的生物碳。而海洋中的1万亿吨生物量只有大约0.1%到0.2%是活的。森林的存在是陆地上活生物量比海洋中多得多的原因。

我们还不清楚地球的**深部生物圈**或地内生命的规模，但对于这个问题的认识将有助于我们理解潜在的地外（例如在火星或木卫二表面以下）生命。一些科学家认为，有大量微生物生活在海床表面以下1到2千米的地方，以及陆地表面以下超过3千米的地方。在这种深度，温度是限制生命生长的一个重要环境因素。越往地下就越热（矿工们都知道这一点）。在某个深度，温度会升得很高，即使是最坚韧的微生物也无法生存繁衍。由于还没有对所有类型的地下环境进行深度钻探，我们目前不确定地球的深部生物圈拥有的生物量，但估计现有的大约占地球总体活生物量的1%到30%。

无论其确切质量如何，地球生物圈还不到地球总质量的十亿分之一，但对地表环境的化学特征产生了巨大影响。生物圈得以做到这一点，在于其能迅速周转极大量的物质。在这个过程中，生物个体的生老病死在地质时间尺度下几乎是白驹过隙。微生物通常在几十分钟到几天内繁殖，而大型多细胞生物最长只能存活几千年，最终成为微生物降解过程需要的"死饲料"。例如，在多细胞生物中，已知最古老的非克隆生物是加利福尼亚州白山上的一棵狐尾松，它萌芽于公元前3049年左右，甚至比最早的埃及金字塔还早了几个世纪，但这在地质历史中只是一个小点。

当今生物圈的主要活动是产氧光合作用及它的逆向反应，

即有氧呼吸和氧化反应（详见第四章）。但是我们稍后就要讨论到微生物还会进行多种多样的其他代谢活动。

显而易见的是，我们可以在地表几乎任何一个地方找到微生物。一种海水和淡水细菌——遍在远洋杆菌——可能是地球上数量最多的生物。尽管如此，它在2002年才被人类首次描述，这也说明生物学仍然有待发展。总的来说，海洋中生活的各种各样的微生物总计约10^{29}个，远远超过可观测宇宙中的10^{22}颗恒星。陆地上的微生物也十分丰富。每克表层土壤通常含有约1亿到100亿个微生物。每立方米室内空气通常含有大约100万个细菌。在海拔32千米的平流层中，平均每55立方米的空气中飘浮着一个微生物（尽管是死的）。

生命得以在地球上如此广泛分布，取决于四个关键环境特性。最重要的是在地表广泛存在的液态水。所有进行代谢的生物都依赖水溶液来分散有机分子。所以，我们不指望在像金星那样完全干燥的环境中还会存在生命。第二个必需的特性是新陈代谢所需的能量。虽然阳光是地球生物圈的主要能量来源，但在黑暗环境中也有一些生物可以通过化学反应获得本身所需的能量。因此，在火星或木卫二表面下也不是不可能存在微生物类型的生命体的。第三个维持生命的特性是基本化学元素的可再生供应。如果一个星球不能通过自然循环（如水循环或构造运动）实现重要元素再补给，那么它将会毫无生机。第四个生命可能必需的环境特性是存在固体、液体和气体共存的界面。对于生物来说，生活在陆地或海洋表面等稳定的物相界面上是非常有利的。这也是生命很难在木星这样没有稳定表面的气态巨行星上存在的原因。据推测，木星大气层的特定高度可能适合生命存在，但由于木星大气存在对流的剧烈搅动，潜在的生命会周期性地陷入致命的高温和高压之中。

地球生命的内部视图：细胞

我们刚刚从全球尺度思考了地球上的生物，但是在显微镜下观察，所有生物均由细胞构成①，并且根据细胞的不同类型，全球生物可以划分为三个域：**真核生物域**、**古菌域**和**细菌域**。后两者是微生物，有时合称为**原核生物**。不过，现在许多微生物学家认为"原核生物"这个术语已经过时了，因为古菌和细菌在生物化学组成上并不相似。在古菌和细菌中，DNA自由漂浮在细胞中，而在真核生物中，DNA则处于有膜结构的细胞核中。古菌和细菌都是单个的细胞，只有一些种类的细菌除外，它们可以排成一列形成菌丝。真核生物也可以是单细胞生物，例如阿米巴原虫或面包酵母等。但只有真核细胞才能形成大型的、具有三维结构的多细胞生物，例如蘑菇或人等。

生物的三域分类是基于遗传学划分的，取代了之前的五界生物分类法，即植物、动物、真菌、原生生物（单细胞真核生物）和细菌。然而，在生物分类学中，这些古老的术语仍在使用。它们将生物在域以下按照界（Kingdom）、门（Phylum）、纲（Class）、目（Order）、科（Family）、属（Genus）、种（Species）来分类。我个人对这些分类单位的助记方法实在是粗糙原始，所以就不提了；但另一个口诀是"帮脾气暴躁的科学家把珍贵的生物管理得井井有条"（"Keeping Precious Creatures Organized For Grumpy

---

① 病毒不具有细胞结构。

Scientists"）。举个例子，人类的分类级别是动物界、脊索动物门、哺乳动物纲、灵长目、人科、人（Homo）属和智人（Sapiens）种。瑞典植物学家卡洛斯·林奈（1707—1778）提出了**生物学**一词，并发明了现代分类学，包括对生物的二名命名法，如"智人"（*Homo sapiens*）。但是，直到进化论和分子生物学出现，人们才得以揭示生命的生物化学统一性和遗传上的共同祖先。

虽然真核生物域和其他两个生物域之间有一些相似之处，但它们在复杂性上仍然有天壤之别。细菌和古菌的大小通常在0.2到5微米（1米的百万分之一），极少有例外，而真核细胞通常更大，为10到100微米。更大的体积使真核细胞可以容纳可实现特定功能的各种细胞器，就像人体的器官一样（图4）。例如，细胞中的**线粒体**具有进行呼吸作用的功能。在植物或藻类细胞中，**叶绿体**进行光合作用。不过，所有细胞的一个共同特征是有大量核糖体。核糖体是制造蛋白质的球状结构。例如，在原核生物或简单的真核生物（如酵母）细胞中，可能有几千个核糖体，而在动物细胞中，这个数字可能达到几百万。

考虑到天体生物学研究的一个方向就是复杂地外生命，我们可能会问，为什么只有真核细胞才能产生大型的、具有三维结构的多细胞生命。尽管我们还不是很清楚这个问题的答案，但相较于古菌和细菌，真核细胞内部有更复杂的动态细胞骨骼或者叫细胞骨架。这一结构由蛋白质微丝、微小的蛋白质管（"微管"）和分子马达组成。这些分子控制着细胞结构，并且帮助运输信号分子，从而调控细胞生理机能。因此，真核细胞的组成决定了它具有发展成（皮肤细胞或脑细胞等）许多特异形式的能力。虽然蓝细菌也能够将上百个不同种类的细胞排成一列组

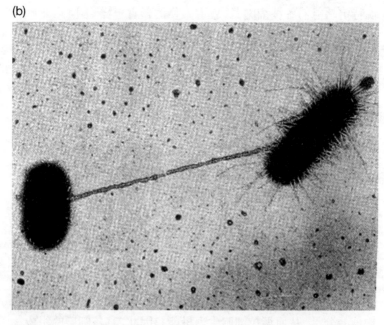

图4 （a）原核细胞（古菌和细菌）与真核细胞结构示意图。（b）两个正在接合的细菌

细丝结构,但这是它的极限。另外一个与古菌和细菌不同的地方在于,真核生物有更大、更模块化的基因组,因此它们可以更加复杂多样。如果没有演化出真核细胞,地球将会变得沉闷许多。我们在现今世界中所熟知的生物,包括动物、植物和真菌,都将不复存在。所以,当我们思考系外行星上是否有复杂生命时,我们应该琢磨其他世界中是否能演化出真核细胞。也是考虑到这个原因,人们对于真核生物的起源有极大的兴趣。

现代遗传学认为,真核细胞是在生物演化历史上由细菌和古菌的碎片组装而成的"弗兰肯斯坦怪物"。例如,真核细胞中的线粒体来源于一个最初与另外一个细胞共生的细菌。这个过程可能是:一个较大的细胞——可能是一种已不复存在的古菌或某种"前真核生物"——吞噬了一个自由生存的细菌,而这一历史上最重要的吞噬创造了线粒体的祖先。事实上,线粒体中的DNA不同于细胞核中的DNA,这为这一"细菌祖先"假说提供了证据。此外,植物和藻类细胞中叶绿体的DNA也不同于细胞核中的DNA,这表明叶绿体可能最开始是一个共生的蓝细菌,后来以类似的方式被吞噬,最终生活在更大的细胞内。实际上,绿色植物叶子中的所有细胞都包含了蓝细菌的祖先。在很久以前,这些蓝细菌"奴隶"就被关在细胞"笼子"里,然后被驯化了。上述提到的真核细胞中线粒体、叶绿体和其他细胞器的起源理论被称为**内共生**。

一个没有真核生物的世界也会是一个没有性的世界。想象一个没有鲜花,也没有情歌的世界。古菌和细菌是没有性别的,但是当两个细胞通过一根管子连接时,它们就会接合,并转移被称为质粒的DNA片段,从而实现基因的传递。细菌表面有一种叫作菌毛的突起。在细胞接合的过程中，一种特殊的菌毛伸长到另一个细菌上，作为基因传递的导管（图4）。与真核生物的有性繁殖不同，微生物接合不产生后代而是改变自身性状，这种方式非常快速、简单。这就好比你在咖啡店里轻轻碰了一个红头发的人，获得了红头发基因，然后你的头发瞬间变成了红色。此外，古菌和细菌也可以通过遗传转化（例如，从环境中吸收外来DNA）和转导（例如，病毒介导下的基因交换）来获得新的基因。事实上，这种快速获得基因的能力恰恰是细菌快速提高耐药性的原因。

相较之下，有性别的真核生物会产生**配子**，即精子和卵细胞。对于复杂的多细胞真核生物来说，生命的多样性让如何去定义"雄性"与"雌性"变得异常困难。我们只能选择一个奇怪的定义：产生小配子的是雄性，而雌性则产生大配子。配子相互结合，因而一半基因来自父亲，另一半来自母亲。通常，DNA是一个松散盘绕的线状物，但在细胞分裂前，可以在显微镜下观察到DNA卷曲成**染色体**。举个例子，人类的每个细胞有23对共46条染色体，只有**配子**除外；配子只有体细胞的一半，也就是23条染色体。

为什么有性繁殖对真核生物的演化是有益的？人们对此众说纷纭。一种假说认为，有性繁殖可以将父母双方的基因混合并匹配到每个染色体上，这个过程也被称为基因**重组**。如果有益的基因突变在两个个体中分别单独发生，那么它们通过无性繁殖得到的下一代将无法同时获得所有的基因突变性状。但是，有性繁殖可以将有益的基因突变集中传递给下一代，从而在演化上更有优势。相反，有性繁殖也可以通过将一些个体的未突变基因结合在一起,来消除那些突变后对生物不利的基因,而自我克隆的生物体则一直携带坏基因,这会导致其后代死亡。

在三域生物分类系统之外,病毒代表了生命与非生命之间的灰色地带。海水或土壤中的病毒数量通常比其他微生物还要多十倍。它们由蛋白质及被蛋白质包围的DNA或RNA片段组成,在某些情况下,还会有膜结构。病毒非常微小,只有50到450纳米(1米的十亿分之一)大小,与紫外光的波长相差无几。在通常的情况下,它们被认为是非生命,因为它们要繁殖就必须去感染和劫持细胞,一旦离开细胞就了无生机。然而,有些病毒可以在几乎不影响宿主的情况下完成这些过程,所以并不是所有的病毒都会引起疾病。关于真核细胞中细胞核的起源,其中一个理论是,它可能是从一个大型DNA病毒演化而来的。不过,病毒在生命演化中的作用仍然是一个有争议的话题。

生命的化学组成

在讨论遗传和新陈代谢等生命的许多特征前,我们需要了解一些生物化学名词。构成生命的分子一共有四大类:核酸、糖类、蛋白质,以及脂类。就像日常用的自组装家具一样,许多生物分子也是模块化的。它们都是由更小的单元——单体——组成的链或聚合物。

糖类可以为生物提供能量并且参与形成生物结构。它们有1∶2∶1的C、H、O元素比,可以看成是$C(H_2O)$单元的重复组合,因此在字面意义上,它们的化学组成就是"碳水化合物"。含有五个碳原子的糖存在于DNA和RNA分子中,而含有六个碳原子的糖存在于细胞壁中,如植物中的纤维素。

脂类是一种不溶于水但溶于非极性有机溶剂，例如橄榄油，的有机分子。它的任何一个原子都不显著带电。因此，如果我们把死去的动物碾碎，浸泡在非极性溶剂中，所有可以溶解的东西都是脂类。生物利用脂类构建细胞膜，将脂类作为储存能量的脂肪或者作为信号分子。细胞膜的主要成分是磷脂，由一个含磷的亲水端和一个碳、氢原子组成的疏水烃链构成。由两层磷脂组成的**双分子膜**构成了细胞膜。其中，亲水端伸入细胞内与细胞外的水介质中，而疏水端则在膜的中间聚集。

蛋白质是由氨基酸单体组成的聚合物。它们的用途主要是形成酶和结构分子，不过还有一长串其他功能。

RNA和DNA都是核酸，是**核苷酸**单体的聚合物。每个核苷酸都由一个五碳糖、一个磷酸基团和一个碱基组成（图5）。在DNA中，有四种可能的碱基。它们都包含一个或两个由六个原子组成的环，其中四个原子是碳原子，两个是氮原子。每个碱基都有各自的字母名称：A、C、G、T，分别代表腺嘌呤（adenine）、胞嘧啶（cytosine）、鸟嘌呤（guanine）和胸腺嘧啶（thymine）分子。RNA分子使用其中相同的三个碱基，只是DNA中的T被尿嘧啶U（uracil）替代了。

1953年，詹姆斯·沃森和弗朗西斯·克里克非常成功地推导出了DNA的结构：两条多核苷酸链呈螺旋状盘绕（图5）。两条链上的碱基通过氢键连接在一起，其中一条链上的碱基上带少量正电荷的氢原子被相对链上的碱基上带少量负电荷的原子吸引。只有C-G配对和A-T配对才能实现结构上的兼容。一个DNA分子有数百万个核苷酸，而细胞中每个染色体包含一个DNA双螺旋。与之相反，RNA分子大多是单链的。然而，如

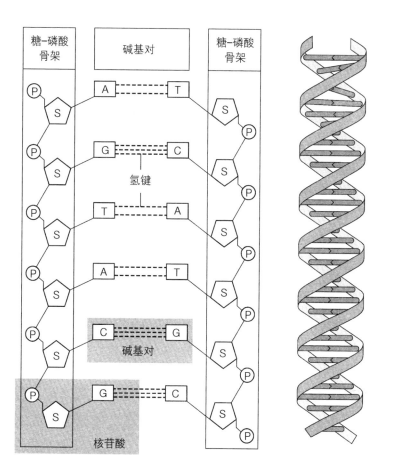

图5 左:DNA由两条相连的链组成,每条链的"骨架"都由磷酸(P)和糖(S)组成。这两条链通过碱基对连接。右:在三维空间中,每条链都是一个螺旋,因此DNA总体上是"双螺旋"

果互补的碱基存在于链的两个独立部分,那么RNA可以发生自身折叠,只是不同于DNA的是,RNA中的腺嘌呤(A)与尿嘧啶(U)配对。

DNA的结构体现了第一章中提到的生命的两个基本特征:自我复制的能力以及为发育和维护提供的蓝图。DNA螺旋在复制时会分裂成两条链，每一条都会作为模板来合成一条新的互补链。比如说，每当模板上出现一个A，新合成的链上就会出现一个T，反之亦然。同样的道理也适用于G–C碱基对。在这个过程中，基因突变和复制错误则为生物演化提供了机遇。

在发挥其另外一个作为蓝图的作用时，DNA双螺旋结构可以被酶解开，以提供合成蛋白质的指令。解螺旋后的部分DNA片段通过**转录**产生信使RNA（mRNA）。信使RNA是DNA的互补拷贝，只是在DNA中出现A的地方对应插入U（而不是T）。随后信使RNA像一条缎带一样被送入核糖体。在传递**遗传密码**的过程中，信使RNA链上的三个碱基的组合（也被称为**密码子**）明确指示了核糖体中合成的蛋白质的每个氨基酸。

除了繁殖，生命还必须通过新陈代谢来维持自身生存，这包括分解分子来产生能量（**分解代谢**）以及构建生物分子（**合成代谢**）。新陈代谢方式的分类取决于能量和碳的来源。由于能量和碳的来源各有两种，因此生物学家将生物代谢类型分为$2 \times 2 = 4$种，分别是**化能异养生物**、**化能自养生物**、**光能异养生物**和**光能自养生物**（图6）。所有的生物（也许甚至包括地外生物）都属于上述一个或多个代谢类型。"**养**"意为供养、摄食；生物获得能量的两种方式——化学能或光能——则决定了代谢类型是"**化能**"还是"**光合**"。"**异**"与"**自**"体现的是生物获得碳的方式：如果碳是通过消耗有机碳成分（如糖）而获得的，则使用"**异**"；如果生命利用无机碳（如二氧化碳）合成有机碳，也被称为碳的固定，则使用"**自**"。一般来说，异养生物必须通过获取食物来获得能量，而自养生物可以固定碳并自己产生能量。

我们人类是化能异养生物。你我所消耗的有机物是由其他

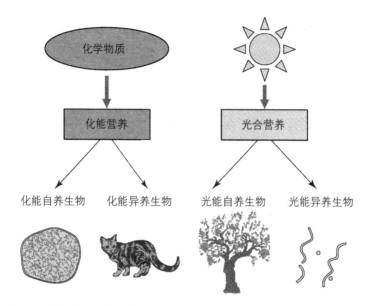

图6 地球生物的代谢分类法

有机体，比如植物产生的。所有动物和真菌、许多原生生物，以及大多数已知的微生物都是化能异养生物。相反，化能自养生物是能够利用氢、硫化氢、铁或氨等无机物产生能量的微生物，其中一些能量被它们用于固定二氧化碳中的碳。例如，有一些化能自养型的微生物生活在黑暗无光的深海热液喷口附近，通过氧化硫化氢和其他物质来获取能量。[化能自养生物有时也被称为chemolithotrophs，这来源于希腊语lithos（岩石），因为它们利用的无机化学物质来自地质资源。]植物、藻类和一些蓝细菌都是光能自养生物，因为它们利用阳光来获取能量，并从空气中获取碳。在热泉中发现的**绿弯菌**属于光能异养生物，它们利用阳光获取能量并从其他微生物产生的有机化合物中获取碳。化能自养生物可能生活在火星或木卫二的地表下面，而光能自养生物可能存在于宜居系外行星的海洋中。

生命之树（网）

我们可以通过研究演化改变基因的方式或速率来推断地球上生命的演化历史。演化是指种群中的可遗传性状从上一代到下一代的变化特征。由于个体的基因是可变的，在任何给定的环境中，都有些个体会比其他个体适应得更好，繁殖成功率更高，生物学家称其具有更高的**适应力**。每一代中适应力更低的个体会被淘汰，这便是自然选择。因此，经过许多代，谱系积累了遗传适应性，新物种便逐渐演化出来。

对有性繁殖生物而言，**物种**是在自然条件下不能杂交产生可育后代的不同群体，例如人类和马。许多细菌和古菌生活在不同的生态位上，依据这些生态位可以定义不同的群体，但微生物之间基因交流的便利使得确定微生物的种类变得更加困难。因此我们需要使用遗传学的手段。负责编码核糖体中的RNA（核糖体RNA）的基因中大约2%到3%的差别就足以作为分辨两个不同微生物物种的依据。

实际上，我们可以通过对比基因来评估所有生命形式之间的亲缘关系。例如，基因表明，同为真核生物的你和你脚趾间的真菌之间的亲缘关系，比古菌和细菌之间要亲近得多，尽管看起来绝非如此。从20世纪60年代中期开始，在分子水平上了解生命的各种技术已经逐渐发展起来，其中包括比较蛋白质中的氨基酸序列或RNA和DNA中的核苷酸序列。

基因决定了蛋白质中的氨基酸序列，因此"蛋白质序列"的差异指示了物种间的明显差异或者亲缘关系。例如，"细胞色素c"

是一种含有104个氨基酸的呼吸蛋白,在许多生物中都发现了它。这一序列在人类和黑猩猩身上是一模一样的,显示出两个物种之间的密切关系。与人类的细胞色素c蛋白质序列相比,恒河猴有一种不同的氨基酸,狗有十三种不同的氨基酸,诸如此类。物种亲缘关系越远,蛋白质序列差异越大。有了这些数据,分析人员就可以画出一棵树,就像家谱树一样,把所有物种联系起来。

20世纪70年代,美国微生物学家卡尔·乌斯(1928—2012)使用核酸而不是蛋白质来确定生物之间的关系。他分析了核糖体RNA序列,敏锐地意识到这种方法可以用于解析所有生命。核糖体的工作——合成蛋白质——是一切细胞的核心功能。因此,编码核糖体RNA的基因会随着时间的推移而仅仅缓慢突变,因为这一过程中的大多数突变都是致命的,无法积累。核糖体RNA占了核糖体干重的约60%,其余则是蛋白质。虽然原核细胞的核糖体比真核细胞的要小一些,但除此之外,它们有类似的结构和功能,因此可以进行比较。

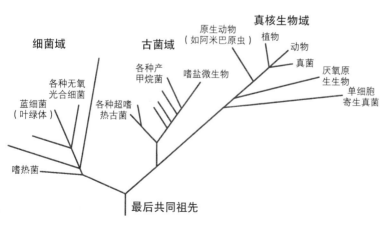

图7 基于核糖体RNA构建的生命之树

乌斯分离出一小段核糖体RNA，并通过对比不同物种间的核苷酸序列的差异构建了一棵"生命之树"（图7）。以这种方式构建生物的演化历史被称为**系统发育**。乌斯的生命之树轰动一时，因为生命被分成了上面提到的三个域，与将古菌和细菌归到一起的传统五界范式完全不同。此外，乌斯的生命之树还将植物、动物和真菌降级至真核生物分支末端的细枝。

如今，虽然乌斯的三域概念已被广泛接受，但对于生物体内其他基因的研究表明，从上一代到下一代垂直下降的树过于简单了。这是由于微生物会随意交换基因（图4），有时甚至会与不相关的物种交换，这被称为**侧向**（或**水平**）**基因转移**。因此，生命之树的诸多微生物分支更像是一张生命之网——分支间由于水平基因转移而变得纵横交错。

自20世纪90年代中期以来，DNA测序已经变得更加自动化。如果你想从环境样本中获取单个基因的序列，你只需要按照如下流程操作：从细胞中分离出DNA→使用聚合酶链式反应多次复制（或扩增）DNA基因→获取基因序列→与其他生物相比较→生成系统发育树。越来越多的人开始对物种的完整基因组进行测序，其中包括包含大约21 000个蛋白质编码基因的人类基因组。令人震惊的是，人类蛋白质编码基因仅覆盖了人类总体30亿个核苷酸的1.5%。其余的序列被称为"非编码DNA"或"垃圾DNA"。这些术语实际上是使用不当的，因为非编码DNA的许多片段调节着某些特定基因的表达（即合成蛋白质）或编码产生非蛋白质产物（如核糖体RNA）。

基因序列分析揭示了古菌与细菌间有令人惊讶的分歧，这也在生物化学成分分析方面得到了证实。例如，细菌的细胞壁含有肽聚糖，它由通过蛋白质交叉连接的碳水化合物棒状物组成，而古菌的细胞壁主要是化学组成不同的蛋白质或碳水化合物，或两者都有。相较之下，真核细胞的细胞壁对于植物来说主要由纤维素构成，对于真菌来说由几丁质构成，而动物细胞没有细胞壁，只有一层细胞膜。此外，细菌和古菌的细胞膜中脂类也不同。首先，两类微生物中，细胞膜磷脂的疏水端和亲水端之间的化学连接不同。其次，古菌的脂类双分子膜的疏水端不是"面对面"地悬挂在层中，而是由分子从头到尾地连接在一起。这增强了古菌的细胞膜强度，也是一些被称作超嗜热微生物的古菌能够在温度很高的热水中生活的原因之一。除了上述成分上的差异，细菌和古菌也有生态学上的差别。例如，古菌均不是病原体，因此你可能会感染细菌性疾病，但绝不会因为接触古菌而染病。

系统发育的一个进一步用途是，类群之间的遗传差异可能与它们在地质历史上分化的时间有关，由此可以制作一个**分子钟**。20世纪60年代，埃米尔·祖克坎德尔和诺贝尔奖得主、化学家莱纳斯·鲍林比较了从由化石证据证明的共同祖先分化出的一些类群的蛋白质。他们发现，氨基酸差异的数目与演化所需的时间成正比。对于这种现象的一种解释是，大多数的突变是中性的，对个体的适应力没有影响，因此主要受时间导致的积累效应控制。类似地，给定DNA序列中的核苷酸替换数量也与演化所需时间成正比。分子钟在分析物种关系密切的类群时效果最好，可能是由于这些类群具有相近的基因突变速率。亲缘关系较远的物种有不同的世代时间和代谢率，因此必须考虑到基因突变速率的变化。在使用分子钟分析时，利用计算机算法拟合遗传分子差异和生物化石的时间记录会得到一个校准点，由此可以确定这些类群的祖先出现的时间。当我们回溯到很久远的地质时期时,化石记录变得越来越少,使得这项技术的应用变得十分具有挑战性。尽管如此,分子钟表明,动物的最后共同祖先出现在距今大约7.5亿至8亿年。奇怪的是,这个计算得到的时间比目前发现的最古老的动物化石要早。

尽管横向基因转移现象使得上述分子钟分析变得更复杂,生命之树还是提供了不少有关早期生命的信息。例如,在80℃至110℃生长得最好的嗜热菌(一类在高温下茁壮生长的微生物)处于古菌和细菌的核糖体RNA树的根部附近(图7)。一个合理的推论是,所有生命的最后共同祖先可能生活在热液环境中。靠近树根部的生命也是化能自养生物,这表明原始微生物可能从无机物中获得所需的能量。这棵树还表明,复杂的生物,如植物、真菌和动物,在较晚时候才演化出来,与从化石记录得到的观察结果一致。因此,如果地球生命的系统发育史是天体生物学的一份指南,它便提供了与化石记录相同的启示意义:其他星球上如果有生命的话,那么它们在大部分历史时期都应该是以微生物的形式存在的。

极端环境中的生命

除了在生命之树上靠近地球生命的最后共同祖先,嗜热微生物也是研究**极端环境微生物**的起点。极端环境微生物是生活在对人类来说条件非常极端的环境中的微生物的统称。1965年,美国微生物学家托马斯·布罗克在黄石国家公园一眼热气腾腾的热泉中发现了生活在82℃到88℃的细菌的粉色菌丝。当时,人们还没有发现任何一种生物可以在73℃以上生存,因此布罗克的发现激发了人们对探索生命极限的兴趣。

不过出人意料的是,布罗克的工作最终也促成了遗传学研究的大爆发。布罗克在另一眼热泉中发现了一种新细菌——水生栖热菌。工业科学家从这种微生物中分离出一种在高温下稳定存在的酶,这种酶能够催化聚合酶链式反应(PCR)。这项DNA复制技术的发明为生物学带来了革命性的巨变。因此,作为这个案例中的纯科学研究,天体生物学最终以一种出人意料的方式造福了社会。遗憾的是,把自己发现的细菌奉献出去的布罗克没有从中得到一分钱。

通常来说,极端环境微生物的生长(即复制)依赖极端环境,或者说它们在那种环境中生长得最好。就生存温度而言,真核生物的上限是62℃,细菌为95℃,古菌为122℃。目前,最高生存温度纪录的保持者是产甲烷古菌或甲烷嗜热菌,它的最佳生长温度是98℃。

如今,人们发现了各种各样的极端环境微生物(知识匣),发现生命存在于多种极端环境中,远远超过了50年前任何人认为的生命极限。这为地外生命的存在提供了可能性。例如,考虑到地球的地壳内部有嗜热菌,因此类似的微生物也可能存在于火星地壳深处。此外,沃斯托克湖(250千米长,50千米宽,1.2千米深)位于南极洲东部厚达4公里的冰层下。这片冰湖上方约100米的冰层被认为是湖水冻结所形成的。奇怪的是,这些冰中含有化能自养嗜热菌的DNA片段。这表明,在寒冷的湖水(由于压力较高,湖水的温度可能只有−2℃)之下可能有热液活动,将嗜热微生物从岩石裂缝中喷发出来。这些为沃斯托克之类的冰湖设计的探测生命的技术,可以应用于地外生命的探索计划。例如，在木星的冰卫星木卫二冰层下的海洋中采集生命样本，也会面临类似的技术挑战。

> **知识匣 一些极端环境微生物**
>
> **嗜酸微生物**：需要pH值为3或更酸的介质才能生长；有些可以耐受pH值低于0的极酸环境。
>
> **嗜碱微生物**：最适宜生长在pH值为9及以上的碱性介质中，有些可以在pH值高达12的环境中存活。
>
> **嗜压微生物**：在高压环境中达到最佳生长状态；一些物种可以在1 000倍以上地球大气压下生存。
>
> **岩内微生物**：生活在岩石的孔隙中。
>
> **嗜盐微生物**：需要盐度较高的介质，至少达到海水的三分之一。
>
> **岩底微生物**：生活在岩石底部，如岩石与土壤的交界处。
>
> **嗜冷微生物**：在低于15℃的条件下生长得最好，有些可以在-15℃的条件下生长，也曾有报道说一些微生物能在低至-35℃的条件下维持代谢。
>
> **抗辐射微生物**：能抵抗如来自放射性物质的电离辐射。
>
> **嗜热微生物**：在60℃到80℃生长旺盛，而**超嗜热微生物**在80℃以上生长最佳；相比之下，生活在15℃到50℃的人类等生物被称为**嗜温生物**。
>
> **嗜旱微生物**：生活在几乎没有生物可利用的水的环境中，可能同时又是嗜盐微生物或岩内微生物。

第六章

太阳系里的生命

现今哪些星球是宜居的？

2002年，在讲授天体生物学课程时，我设立了一个奖项，奖励能猜出冥王星轨道以内我认为可能存在地外生命的九个天体的学生（表1）。当年，没有学生拿到这个奖。但是，随着对天体生物学认识的提高以及网上信息的增多，有学生在2010年获得了该奖项。今天，我可能会再加上几个天体，但让我们把参考答案留到本章的最后来揭晓。

我在表1中考虑的潜在地外生命类型是很简单的、类似于微生物的生物，主要的判断标准是液态水的存在。在地球上，只要有液态水的地方，我们都能发现生命，无论是在咕嘟冒泡的热泉中，在冰里的卤水包裹体中，还是在地壳深处包裹着矿物的水膜中。

有些人的思维更发散，推测在土星最大的卫星土卫六上的湖泊中，可能存在不依赖水的**怪异生命**。土卫六上的湖泊含有液态碳氢化合物，而不是水，有点像小规模的石油海洋。目前，人们还不清楚是否存在可以单纯使用碳氢化合物作为溶剂的生命。我稍后会提到,有一些人基于物理化学的基本原理,认为可能很难出现这样的生命。相比之下,液态水可以支持生命是毫无疑问的。

表1 现代太阳系中九个可能存在生命的天体。这些天体到太阳的距离用天文单位(AU)来表示;1 AU代表地球到太阳的距离,约等于1.5亿千米

天体	天体类型与到太阳的平均距离	为什么可能存在生命
火星	行星,1.5 AU	地表下可能存在液态水
谷神星	太阳系中最大的小行星,2.8 AU	可能存在冰下海洋
木卫二、木卫三、木卫四	木星的大型冰卫星,5.2 AU	有证据表明存在冰下海洋
土卫二	土星的冰卫星,9.6 AU	有证据表明存在冰下海洋和有机物
土卫六	土星的最大卫星,9.6 AU	有证据表明存在冰下海洋和有机物
海卫一	海王星的最大卫星,30.1 AU	可能存在冰下海洋
冥王星	大型柯伊伯带天体,39.3 AU	可能存在冰下海洋

阳光和太阳系内行星的宜居性

太阳系的内行星,包括水星、金星、地球和火星,是否有液态水主要与表面温度有关,也就是与到太阳的距离有关(表2)。"位置,位置,位置"不仅仅是卖房子的时候要考虑的关键因素,对行星的宜居性也至关重要。

位置之所以那么重要,是因为阳光呈球面传播,其表面积随行星—太阳距离的平方而增长。**太阳辐射通量**则是每平方米接

表2 太阳系中内行星和影响它们现今宜居性的因素

	水星	金星	地球	火星
至太阳的平均距离（AU）	0.4	0.72	1	1.5
平均表面温度（℃）	167	462	15	−56
温室效应（℃）	0	507	33	7
大气压强（kPa）	0	9 300	100	0.6
是否有液态水？	太热	太热	丰富	大多数情况下太冷
大气主要成分	几乎没有大气	96.5%二氧化碳 3.5%氮气	78%氮气 21%氧气	95.3%二氧化碳 2.7%氮气

收阳光的瓦数,好比一个灯泡的功率。在地球轨道处,也就是距太阳1个天文单位的地方,太阳光在每平方米面积上提供1 366瓦的能量,几乎相当于14个100瓦的灯泡。在木星轨道处(距太阳5个天文单位),单位面积接收到的太阳辐射通量为地球的1/25,因为同样多的太阳出射能量会散布在面积为5×5=25倍的光球面上。在火星轨道处(距太阳1.5个天文单位),单位面积接收到的太阳辐射通量为地球的1/2.25。相比之下,在距太阳0.72个天文单位的金星轨道处,单位面积接收到的太阳辐射通量几乎是地球的两倍。

距太阳只有0.4个天文单位的水星显得毫无生机。它是八颗行星中最小的,直径大约只有地球的五分之二。它表面没有液态水,也可能从来没有过。如今,水星荒芜的表面有时能达到430℃。考虑到水星引力小,又受到强烈的阳光照射,因此即使水星上曾经有大气,也会在水星刚形成时就消耗掉了。

金星和火星同太阳之间的距离解释了它们为什么不利于生命存在,尽管这并不是全部原因。金星约460℃的表面温度甚至比水星表面还要高,而火星表面是一片冰冷的沙漠,平均温度为−56℃。金星的大小与地球相当,但火星要小一些,直径大约是地球的一半,而质量是地球的九分之一。事实上,火星正是由于体积较小,才变成了现今不宜居的状态,这一点我们稍后会谈及。

虽然太阳辐射通量是影响行星表面温度的一个因素,但火星和金星上的温室效应也起着关键作用。相比于地球上1 kPa的大气压强,火星大气仅仅产生0.006 kPa的压强。这种稀薄的大气是由95.3%的二氧化碳、2.7%的氮气和少量其他气体组成的干燥的混合气体。回想一下上文,地球大气提供的温室效应是33℃。由于大气稀薄干燥,火星上的温室效应只有7℃,导致其表面处于冰冻状态。金星大气中有两种主要气体,即96.5%的二氧化碳和3.5%的氮,其比例与火星上的气体相似。但与之形成鲜明对比的是,金星的大气层非常厚,平均地面高度处的大气压强高达93 kPa。这就导致金星大气的温室效应高达507℃,比火星高500℃(表2)。所以,两颗行星的表面都不能够支持稳定的液态水存在。在火星稀薄的空气中,一个液态水坑会同时发生沸腾(或迅速蒸发)和冻结,并且火星表面的水通常只以冰或水蒸气的形式存在。另一方面,在灼热的金星表面,液态水也是不可能存在的。

由于没有液态水,大多数天体生物学家认为金星上应该是没有生命的。但是,也有一些人猜测,金星上由硫酸颗粒构成的云中可能生活着嗜酸微生物。然而,我对这种猜测表示怀疑。这是因为金星云层缺乏生物可用的水。此外,金星的大气湍流会把云层中任何潜在的微生物拉到炙热的低处或致命地干燥的高处。

金星过去是否有生命?

虽然现在金星上不大可能有生命存在,但是关于金星的天体生物学研究仍然很有吸引力,因为它过去可能有海洋,甚至是生命。考虑到金星上除水之外的挥发分含量和地球上类似,我们可以推测最早期的金星应该也拥有大量的水。这里解释一下,挥发分是指在正常行星温度条件下可以变成气体的物质。例如,如果你把地球上所有的碳酸盐岩石中的碳转化为二氧化碳,那么地球表面就会有大约90个标准大气压的二氧化碳,和现代金星相似。类似地,如果你从地球上的矿物质中提取出所有的氮,并将它们释放到地球的大气中,你会得到近3个标准大气压的氮气,这也与在金星大气层中观察到的情况一致。考虑到含水小行星的吸积是金星和地球上碳和氮的最终来源,我们有理由推断,金星像地球一样从小行星那里获得了大量的水。

遗憾的是,金星太过靠近太阳,所以表面发生了**失控的温室效应**,它注定变成现在的样子。当早期金星的表面被强烈的太阳光烘烤时,水的蒸发会使其大气中富含水蒸气。这些水蒸气完全阻挡了从金星表面发射的红外辐射。此外,行星表面通过向外发射红外辐射来降温有一定的功率极限,这主要取决于水的性质和行星的引力。近期,理论计算表明,地球的**失控极限**是282瓦每平方米,而金星还要低一些。

为了理解上述失控极限的概念,我们可以想象一下如何把地球一步步变成现在的金星。我们前面曾经提到,地球轨道接收的太阳入射能量通量为1 366瓦每平方米,但地球表面反射了其中的30%,因此只吸收了70%。加之由于只有半个地球处于日光中,太阳入射总能量通量又减少了50%。此外,鉴于地球弯曲的表面会导致更多的光从地球表面和大气中掠过被反射回太空,入射总能量还需要考虑约50%的损失。把所有因素都考虑在内后,地球平均每平方米净吸收($0.7 \times 0.5 \times 0.5$)$\times 1\,366 \approx 240$瓦的太阳光能量。在一个稳定的状态下,地球会以红外线的形式向太空发射等量的能量,从而保持恒定的全球平均温度。但想象一下,如果我们把地球移动到现在金星的轨道上,地表平均每平方米吸收的太阳光能量将从240瓦翻倍到480瓦。在这种情况下,海洋将变成一个热气腾腾的浴缸,并且湿热的大气将达到能量逸散的极限,即每平方米只有282瓦辐射到太空中。因此，每平方米输入的能量（480瓦）将会比输出的能量（282瓦）多，导致地球变得越来越热。长此以往，海洋会完全蒸发掉，表面的岩石会融化。届时，灼热的高层大气发出的近红外光将照向太空，从而使入射的阳光能量和辐射出的能量恢复平衡，表面温度也将稳定在1 200℃左右。我们认为金星上就发生过这种恐怖的过程。因此，即便金星上存在过生命，它们最终也会被烤焦。

这种失控的温室效应还会导致金星上的海洋消失。在高层逸散大气中，紫外线会将水蒸气分解成氢原子和氧原子。氢原子非常轻，因此会快速逃逸到太空中，这个过程伴随着一部分氧原子的逃逸。留下的多余氧会氧化下面较热的岩石。最终岩石会冷却凝固，但此时大气中已经充满了从炽热的岩石表面释放出来的二氧化碳和氮气。这些过程的最终结果就是，今天的金星环境如地狱一般。

在发生失控的温室效应之前，生命可能已经在金星的海洋中起源。还记得我们前面说过早期的太阳不那么明亮吗？所以，可能直到金星诞生几亿年后，失控的温室效应才开始发生。遗憾的是，即使金星上存在过生命，我们也很难证实。20世纪90年代，NASA的"麦哲伦号"探测器利用雷达绘制了金星表面的地图。一般来说，更老的行星表面会有更多的陨石坑，所以科学家可以利用陨石坑的密度来估计行星表面的年龄。人们利用陨石坑计数推算，金星的表面年龄约为6亿至8亿年。金星的表面似乎被岩浆整体翻新过。一个可能的原因是，金星内部产生的热量无法得到有效释放，可能会逐渐积累，直到内部变得非常热，最终产生周期性的全球性岩浆喷发事件。地球与金星不同的是，地球内部产生的热量可以通过产生新的海底而不断释放，而金星没有水来润滑板块构造活动。

这种地表物质的翻新会破坏任何可能存在的化石记录。然而，如果金星上有过生命的话，那表面仍然有可能保留着隐晦的地球化学痕迹。为了研究金星上是否有过生命，最好的办法是在未来的太空任务中采集金星上的岩石样本，并把它们带回地球分析。

有水的火星：地外生命的栖息地？

与金星不同，许多人设想火星可能是地外生命的一个潜在栖息地。在太空时代之前，人们已经知道了火星的许多基本参数，比如火星上一天有24.66个小时、一年有1.9个地球年、引力是地球的40%。但是，人们对于火星表面的情况知之甚少，直到1965年NASA的"水手4号"第一次实现人类火星飞掠探测。"水手4号"发现火星表面像月球一样布满陨石坑，这打击了人们对存在火星生命的信心。随后，在20世纪70年代初，NASA的"水手9号"轨道飞行器拍摄到火星表面存在干涸的河谷和死火山，这表明火星曾经非常像地球。但很快希望的钟摆又摆回了另一个方向。NASA随后发射了"海盗号"，它由两组相同的轨道器和着陆器组成。"海盗号"于1976年到达火星，但没有发现生命信号。这点我们稍后会讨论。

20世纪80年代，对于火星的探索经历了一段停滞，之后才逐渐恢复。人类发射了"火星全球勘测者号"，从1997年到2006年均在火星轨道上运行。它测绘了火星的地图，并拍摄了火星表面沉积层的图像。这些沉积层的发现意味着，火星曾经存在过许多次侵蚀和沉积的地质循环。在21世纪初，人类先后发射了"火星奥德赛号"、"火星快车号"以及"火星勘测轨道器"，发现了可能在液态水环境中形成的黏土矿物和盐类区域。2004年,"火星探测漫游者"计划的两台火星车登陆火星,发现了由过去的液态水形成的波痕化石与各种沉积岩。2008年,NASA的"凤凰号"着陆器在火星极地地区挖出了地下冰,并测量了土壤中的可溶性盐。之后,在2012年,"好奇号"火星车向直径150千米的盖尔陨石坑内一座5千米高、由沉积层构成的山体进发。它发现了过去水沉积形成的泥岩,其中含有生命所需的SPONCH元素(详见第一章)。①

如今,有关火星的天体生物学研究主要考虑的是数十亿年前形成的潜在火星生物痕迹,或是可能在现代火星表层以下艰难生存的类似于微生物的生命。太空时代之前人们对于一个像地球一样的火星环境的期待,如今已经被残酷的现实所取代:寒冷、狂风肆虐、遍布全球的沙漠,那里每天都有沙尘暴,而且没有降雨。

目前火星的地表环境不利于生命存在,原因有三。第一,虽然我们可以肯定冰的存在,但没有液态水存在的确切证据。例如,火星极地冰盖主要是水冰,顶部有二氧化碳凝华形成的干冰。当冬季来临时,火星极地上空约30%的大气会凝华,形成更多的干冰。此外,在火星中高纬度地区,火星表面以下就是含冰土壤或**永久冻土**。在火星的热带地区,下午时土壤上层几厘米的温度会上升到冰点以上,但在那种环境中,冰在达到融化温度之前就变成了水蒸气。第二,火星没有臭氧层,这导致对生物有

---

① 作者在此处新增的内容为:"2020年,人类发射了'毅力号'火星探测车,它于2021年登陆火星,着陆点是直径45千米的杰泽罗陨石坑。那里曾经是一片湖泊,河流带来的沉积物在此沉积下来,因而可能存留三角洲的遗迹。'毅力号'找到了过去流动水体和地下水渗入岩石的痕迹。它采集了一些火星样本并存放在样本管中,等待在未来执行样本返回任务时把它们送回地球。"

害的紫外线可以到达地表。第三，火星大气中的化学反应会产生过氧化氢。这是一种在漂白剂中使用的化学物质。过氧化氢分子可以沉降到火星表面，破坏潜在的有机物。

虽然在火星表面发现生命的前景比较渺茫，但火星地下的地热能可能支持液态水和生命存在。事实上，从2004年开始，有一些报道指出火星大气中存在平均体积占比为一亿分之一的甲烷分子。这些报道让人们兴奋地认为，火星地下可能存在产甲烷菌。天文望远镜和"火星快车号"轨道器通过分析火星表面反射的阳光，发现火星大气似乎有甲烷的光谱吸收信号。然而，这些甲烷的信号非常微弱，几乎无法分辨，这导致一些科学家（包括我）仍然怀疑火星大气中是否真的存在甲烷。后来，"好奇号"火星车没能探测到超过检出限（十亿分之一体积占比）的甲烷信号。①

在讨论潜在的早期火星生命时，天体生物学家参考了火星的地质时期划分。火星的地质时期分为**前诺亚纪**（距今41亿年之前）、**诺亚纪**（距今41亿至37亿年）、**西方纪**（距今37亿至30亿年）和**亚马逊纪**（30亿年前至今）。人们根据地表陨石坑的不同特征将火星表面划分到不同的地质时期。更古老的火星表面积累了更多且更大的陨石坑。事实上，这种利用陨石坑计数的方法来自对月球的研究。"阿波罗号"计划中宇航员带回的月球岩石的年龄是可以通过放射性同位素来测定的。这些通过同位素定年确定的年龄与相对应的月面上陨石坑密度有一个定量关系。火星可能比月球受到更多的天体撞击，不过人们可以通过

---

① 作者将此处更新为："后来，'好奇号'似乎检测出了约百万亿分之四水平的甲烷。但是使用更严苛的千万亿分之二检出限的欧洲航天局（ESA）的痕量气体轨道器并未找到甲烷，因此一些科学家还在怀疑'好奇号'检出的甲烷来自其自身内部的污染。"

天文计算来校正这种差异，从而将从月球研究得到的陨石定年规律应用到火星上。

火星上有各种各样液态水曾经存在的证据，表明火星曾经比现在更适合生命生存。图片显示了河流冲刷形成的**流水**地貌特征，包括冲沟、干涸的河谷、三角洲和开阔的河道。此外，土壤和一些岩石中也含有在液态水中才会形成的矿物。

在火星南北半球纬度30°至70°之间，陨石坑壁和台地岩壁上形成了数十至数百米长的切口，被称为**冲沟**。这些冲沟上面没有更新的陨石坑覆盖，有些甚至还上覆在沙丘表面，所以它们一定是最近形成的。最初，人们认为它们是在冰融化时形成的。然而，一些观测图像显示，二氧化碳霜升华时会形成这些结构，可能是二氧化碳霜的升华导致了干燥的土壤和岩石的移动。

相比而言，火星上一种更为古老的地貌特征是**河谷网络**。这是一种干涸的河流状洼地，有树枝状分支谷道（图8）。大多数谷道都形成在撞击严重的诺亚纪平原上，宽度为1到4千米，深度为50到300米。相比于地球河流网络来说，这些河谷网络的支流密度要低得多，但我们可以根据一些支流的密度推断，早期雨水或融水形成了地表径流。此外，一些有粗短支流的河谷也可能是由**基蚀作用**形成的，基蚀作用就是火星地下（融）水造成上覆地面的侵蚀和塌陷。一些河谷的尽头还发现了三角洲。

火星诺亚纪的地貌特征主要是一些边缘被侵蚀的、浅浅的陨石坑。目前尚不清楚这种侵蚀是如何形成的。火星河谷网络主要横切发育在陨石坑之上，所以它们应该不是形成侵蚀的原因。随着距今37亿年的西方纪开始，陨石坑的侵蚀率急剧下降，河谷网络也变得罕见。

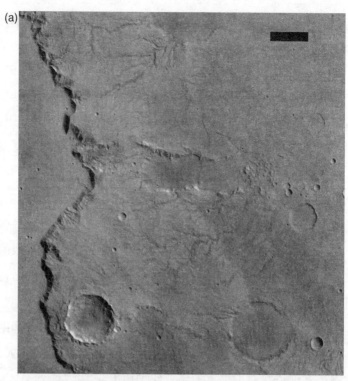

图8 （a）火星上的河谷网络。左侧是直径456千米的惠更斯陨石坑被侵蚀的边缘。图片中心位于南纬14°，东经61°，上为北方。比例尺为20千米。
（b）长约205千米的拉维谷溢流河道，位于南纬0.5°，东经318°

在距今约30亿年的西方纪末期,火星上开始出现**溢流河道**(图8)。与河谷不同,溢流河道是由封闭在堤岸之间的水流动形成的,缺乏支流且来源单一。溢流河道规模很大,有10到400千米宽,长度可达1 000千米左右,深度可达几千米。大多数溢流河道起始于火星地面塌陷形成的**混沌地形**,也有些起源于峡谷或裂口,那里现在有山一样的硫酸盐堆积,可能是盐水蒸发后的残留物。

目前,解释溢流河道的主流假说是,它们是由地下冰融化或蓄水层破裂所形成的洪水导致的。地球上也有类似溢流河道的地貌,例如,美国华盛顿州东部被洪水侵蚀的土地被早期定居者称为**疤地**。这些疤地出现于末次冰期结束时,由一片大型湖泊的冰坝间歇性地决溃形成。相比之下,形成火星溢流河道所需的水量是形成华盛顿州疤地的水量的10到100倍。

要解释侵蚀形成火星溢流河道所需的巨大水量并不容易。据估计,这些水量相当于几百米深的全球性火星海洋,远远多于现存的火星冰储量。由于这些河道流向北部洼地,一些科学家推测在那里形成过海洋。不过,也有少数人认为溢流河道不是由水,而是由岩浆侵蚀形成的。一些化学研究表明,火星岩浆应该较稀,流动性较高,并且具有侵蚀性。

除了上述提到的古老的河谷网络和溢流河道(假定是由水侵蚀形成的),矿物研究也为早期火星更加湿润的表面环境提供了佐证。每个人都知道,古老墓碑上的字是如何通过与水发生化学反应而逐渐消失的。这种反应可以改变或溶解矿物,也被称为**化学风化**。与金星和地球的海底类似,火星的大部分表面是由玄武岩构成的。玄武岩是一种深色的火成岩,富含铁和镁的硅酸盐矿物。玄武岩经化学风化后，会产生**蚀变矿物**，例如黏土。因此，蚀变矿物的存在就意味着液态水的存在，甚至有时还可以指示水的酸碱性。例如，碱性水倾向于使玄武岩风化形成黏土矿物。

　　人们通过分析火星表面发射和反射的红外辐射，已经识别出了多种含水蚀变矿物。但只有约3%的诺亚纪地层表面发现了含水黏土和碳酸盐。硫酸盐矿物在诺亚纪末期或西方纪地层区域大规模出现，而干燥的红色铁氧化物则在年轻的亚马逊纪地层表面很常见。这种发生模式可能指示了三个环境时期。在第一个时期，碱性或中性的地表水风化玄武岩，形成黏土矿物。在第二个时期，火山活动喷出的含硫气体产生了硫酸，形成了大量硫酸盐。第三个时期一直延续至今，火星表面的环境寒冷、干燥，形成了干燥的铁锈色地表。

　　"火星探测漫游者"计划中的一台名为"机遇号"的火星车曾经降落在诺亚纪地层的硫酸盐沉积物附近。"机遇号"火星车发现了毫米级氧化铁——赤铁矿小球——嵌在硫酸盐沉积层中，就像松饼中的蓝莓。这些赤铁矿小球是由37亿年前渗透到火星地下的水中的矿物质沉淀形成的。此外，沉积物中保留的波痕表明，当时的水在地表似乎积累成了深及脚踝的池塘。另一台名为"勇气号"的火星车则在火星的另一侧发现了古代热泉的证据。

早期火星的大气与气候

　　上述一系列表明火星上曾有液态水存在的证据让我们好奇，火星早期的气候是否温暖湿润。科学家莫衷一是，分成两大阵营。一个阵营认为，火星曾有持续数千万年或数亿年的温暖湿润气候。另一些人认为，目前观测到的现象都可以用长期寒冷气候中发生的冰的短暂融化来解释。

第一种情况显然对生命更有利。但遗憾的是，还没有人能够解释早期火星如何才能保持数百万年的温暖状态。约37亿年前，太阳的亮度要比现在弱25%。在这种情况下，火星表面要保持在冰点以上需要大约80℃的温室效应，远高于现代地球33℃的温室效应。人们普遍认为，火星形成后，大气中的氢迅速逃逸到太空中，使大气氧化，充满二氧化碳和氮气。但火星不可能有类似金星的非常厚的大气，因为在火星的轨道位置上，二氧化碳会凝华成冰并形成云。此外，高浓度的二氧化碳大气也会将更多的太阳光散射回太空，从而使火星表面变冷。所以，二氧化碳不能提供足够的温室效应来维系早期火星温暖的气候。另一种说法是，火山喷出的二氧化硫是早期的主要温室气体。然而，二氧化硫溶于雨水，因此当早期火星变得湿润时，大气中的二氧化硫会被降水有效地移除。此外，高层大气反应也会使二氧化硫转化为细粒硫酸盐悬浮颗粒，然后反射更多的太阳光，使火星表面变冷。地球上也发生过类似的事件，例如1991年至1993年，菲律宾皮纳图博火山喷发导致了全球规模的降温。

另一个阵营则提出了多个假说，用来解释如何在寒冷气候下保有液态水。例如，他们提出，外来物质撞击会使冰升华成水蒸气，进而产生降雨，侵蚀火星表面，形成河谷。还有一个假说认为，在早期火星轴倾角和轨道形状的偶然叠加作用下，火星局部地区可以发生冰雪融化，从而形成侵蚀的地貌特征。随着时间的推移，其他行星的引力也会影响这些作用过程。轴倾角是行星的自转轴相对于行星的公转轨道平面的倾斜角。它的存在使得行星上的一个半球比另一个在轨道的特定位置上得到更多的阳光，因此形成了季节的更替。地球的轴倾角是23.5°。与地球不同的是，火星没有一颗较大的卫星来稳定它的旋转轴，而两个小卫星的影响可以忽略不计，所以火星的轴倾角在过去45亿年间一直在0°到80°之间变化。在轴倾角比较大的情况下，夏季时，极地冰将受到更多的太阳照射并升华。气流可以将这些水蒸气输送到寒冷的赤道地区形成降雪。在其他季节或轴倾角较小时，这些雪会在阳光照射下融化而产生融水、造成河流侵蚀。这个假说也可以解释现在观察到的一些存在于中纬度地区的沙尘斑块区域。在过去几百万年里，火星适中的轴倾角可能导致了这些区域中冰层的蒸发，形成了这些沙尘斑块。最后，也有观点认为，火星上的盐水可以在远低于0℃的条件下仍然保持液态。在地球上，我们在冬天时向结冰的道路上撒氯化钠，因为它可以将水的冰点降低到-21℃，使冰融化。然而，"凤凰号"着陆器在火星土壤中检测到的另一种盐——高氯酸盐，它主要以镁盐或钙盐的形式存在——可以将水的冰点降低到-60℃以下。

　　无论有关早期火星的真相是什么，火星较小的体积最终破坏了它的宜居性。小的天体比大的冷却得更快，火星内部热量损失得快，所以广泛的火山活动很久以前就停止了。如果没有火山活动，火星就无法循环补充大气中的二氧化碳，最终使得大气中的大部分二氧化碳转化为碳酸盐。如果早期火星有浓密的大气，那么这么大量的二氧化碳埋藏会形成大量碳酸盐。但是，人们观测到的火星表面碳酸盐并没有那么多。所以，除了上述的碳酸盐沉积，早期火星大气的一部分可能被大型彗星和小行星撞击而逃逸到太空中,这就是所谓的**撞击侵蚀**。由于火星引力较小,它的大气层很脆弱。随着后来更多的大气逐渐逃逸到太空中,火星大气最终变得像现在这样稀薄。

寻找火星生命

尽管人们一直在努力寻找,但我们仍然不确定火星上是否有生命或者有过生命。例如,1976年,"海盗号"着陆器曾经试图探测火星生命痕迹。此外,自20世纪90年代以来,科学家们一直在火星陨石中寻找过去生命的迹象。

每台"海盗号"着陆器都设计了三项实验来检测火星上潜在生命的新陈代谢活动,还设计了第四项实验来寻找有机分子。第一项是**碳同化实验**,用于研究是否有火星微生物可以从空气中获取碳。采集的火星土壤(有些与水混合)被暴露在自地球带来的二氧化碳和一氧化碳气体中,每种气体中都含有放射性同位素碳-14。经过培养,科学家在火星土壤中发现了碳-14的信号。但是在对照实验中,火星土壤经过160℃灭菌处理,却仍然可以吸收碳-14。这些证据表明,碳-14的吸收是由无机化学反应导致的,而非火星微生物活动的结果。第二项测试是**气体交换实验**。该实验监测了火星土壤与从地球带来的有机物溶液的混合物,来分析是否有生物进行新陈代谢产生气体。令人惊奇的是,培养的土壤中出现了氧气,但灭菌处理过的土壤也有释放氧气的信号。显然,这些氧气是由于土壤中的一些化学物质遇水或热分解产生的。第三项研究是**标记释出实验**。向土壤中添加含有碳-14的有机物,如果有火星生物可以利用这些有机物,就会释放出含有碳-14的二氧化碳。结果表明,实验组释放了放射性气体，而经过灭菌处理的土壤没有释放出放射性气体。从表面上看，这是火星存在生命的信号。但大多数科学家认为，这些二氧化碳是由火星土壤中的氧化剂与有机物反应产生的，而灭菌的高温条件使得氧化剂失活了。支持这种解释的一个证据来自第四项实验。在这项实验中，研究人员用一种叫作**气相色谱质谱仪**的设备分析了从加热的火星土壤中飘出的分子。结果表明，在十亿分之一检出限条件下，火星土壤中没有发现有机物信号。

总之，这项探测计划得到的教训是，在寻找地外生命之前，你需要了解目标环境中的无机化学特征，以避免出现假阳性。事实上，"海盗号"计划的生物实验负责人哈罗德·"查克"·克莱因（1921—2001）曾告诉我，在"海盗号"计划构想之初，他只想了解火星土壤的无机化学特征，但NASA负责管理资金的主管们却坚持要直接寻找生命。

20世纪80年代，当科学家们还在仔细琢磨"海盗号"计划的实验结果时，人们取得了一项了不起的发现：地球上已经有来自火星的岩石样本了，那就是火星陨石！天体撞击从火星表面撞掉了一些岩石，其中一部分落到了地球上。事实上，每年大约有50千克火星陨石落在地球上，其中大部分落在海洋里。通过分析一些在撞击弹射导致的轻微熔化过程中密封在陨石中的气体，人们可以证明陨石来自火星，因为这些气体的成分特征与"海盗号"着陆器测量的火星大气一致。对于其他没有密封气体的陨石，我们可以通过分析硅酸盐矿物的三氧同位素（氧-16、氧-17、氧-18）组成，来确定它们是否来自火星，因为所有火星陨石都有独特的三氧同位素特征。截至2013年，人们已经确定了

67颗火星陨石,而这份名单还在不断增加。

1996年,有研究称,在ALH84001陨石中发现了可能的生命遗迹。这颗火星陨石是在1984年("84")对于南极洲的艾伦山("ALH")的一次考察中发现的第一颗("001")火星陨石。ALH84001是一块在41亿年前结晶形成的火星火成岩。岩石内部有一些直径约0.1毫米的碳酸盐球体,形成于40亿至39亿年前。科学家在其中发现了四种可能的生命痕迹。第一种是所谓的微体化石;第二种是据说由微生物活动形成的碳酸盐;第三种是痕量的多环芳烃(PAHs),它是由碳原子组成的六边形环状有机物;第四种是一些磁铁矿(Fe_3O_4)晶体,据说与某些细菌体内的矿物类似。地球上的趋磁细菌可以在它们的细胞内制造磁铁矿晶体,并且在演化中磨砺成了磁铁。这些细菌可以利用自身携带的磁铁矿罗盘顺着地球的磁场线上下移动,以找到下方低氧区和上方含氧区之间的边界,从而进行更高效的新陈代谢。

但在随后的几年里,大家对这四项证据都提出了质疑。例如,有学者指出所谓的火星微体化石只是**看起来像**微生物的棒状结构。此后,科学家们也发现了形状类似的无机矿物。此外,ALH84001中微体结构的体积是地球微生物的十分之一,甚至可能小于生命必需的生物化学过程所需的最小体积。对于第二项证据,流体的蒸发也可以产生球状的碳酸盐沉积,因此生物成因的碳酸盐几乎没有说服力。关于第三项证据,后续的分析表明,大多数检测到的多环芳烃是陨石在掉落到南极后的13 000年间获得的。在地球上,多环芳烃是由有机物质燃烧产生的、普遍存在的大气污染物。由于陨石颜色较暗,能吸收更多的太阳光,所以在夏季时,它们往往矗立在南极冰层上的水坑里,导致水和化学物质渗透到陨石裂缝中。此外,即使ALH84001的内部曾经有来自火星的多环芳烃,它们的成因也有可能是无机的。因为在被抛射出来之前,这块岩石会被撞击加热。这种加热过程会导致碳酸盐热解,释放出含碳的气体,通过进一步与水反应生成与生命活动无关的有机物。第四项证据是关于磁铁矿的。事实证明,ALH84001中只有一小部分磁铁矿具有类似生物的形状。一些科学家认为,许多形状的磁铁矿,包括那些类似细菌的样本,是在陨石被弹出前的热冲击过程中伴随着碳酸铁分解形成的。

关于ALH84001的种种争议表明,要证明在古老的岩石中存在微体生物是多么困难。不过,大多数地球岩石中也缺乏化石,因此即使在ALH84001中没有发现生命痕迹也并不意味着火星上不存在生命。我们需要继续寻找。

谷神星:一个重要的潜在天体?

在火星轨道之外和木星轨道之内,有一个由数百万小型岩石天体组成的小行星带。其中最大的是谷神星,直径约为950千米,距离太阳2.8个天文单位。谷神星的表面含有黏土矿物和水冰。和在火星上一样,黏土矿物表明过去有液态水活动。有关谷神星内部结构的一个主流模型是一个被100千米厚的冰壳包围的岩石内核。在内核上方和冰层下方,可能有一个液态水海洋。它的盐度可能非常高,让它在零下也可以保持液态,并且它也许曾经流到冰壳表面。

生命可能会存在于谷神星海底的热液喷口吗?谷神星是如此之小,以至于现在可能没有什么内部热量支持热液系统,所以即使有生物,它们的生物量也极低。但也许谷神星在过去更宜居。NASA发射的"黎明号"探测器将于2015年抵达谷神星。届时,这个任务收集到的红外发射和反射光谱,应该能给我们提供更多关于表面成分和宜居性的细节。①

木星的冰质伽利略卫星

考虑到第五章给出的气候原因,在小行星带之外的木星似乎不该是天体生物学可行的目标。但木星有67颗卫星②,其中三颗可能适合生命生存。1610年,伽利略发现了木星最大的四颗卫星,分别是木卫一、木卫二、木卫三和木卫四(图9),它们依次离木星越来越远。木卫一和木卫二的大小与月球相似,而木卫三和木卫四的体积与水星相当。靠外的三颗卫星内部是岩石,表面覆盖着冰,而木卫一从内到外都是岩石。木卫一、木卫二和木卫三可能也有铁质的核。

图9 木星的伽利略卫星:木卫一、木卫二、木卫三、木卫四

---

① "黎明号"已于2015年3月进入谷神星轨道,并于2018年11月因燃料耗尽而退役。作者将此处更新为:"'黎明号'在2015年至2018年对谷神星进行了研究。红外发射和可见光反射图片与光谱揭示了谷神星表面的很多细节与宜居性特征。例如,谷神星冰面存在几处主要由碳酸钠组成的明亮斑点,可能是冰下的盐水渗透出来后干涸形成的沉积信号。"

② 截至2024年12月,已经发现了95颗木星的天然卫星。

木卫一是太阳系中火山活动最剧烈的天体。这些火山活动喷出的二氧化硫会在木卫一的表面冻结。但木卫一上没有液态水，肯定没有生命活动。木卫一上之所以有这么剧烈的火山活动，是因为木星对木卫一的引力随着木卫一在椭圆轨道上的运行而变化，就像人挤压一个压力球一样在挤压木卫一。对木卫一的这种**潮汐加热**是由固体物质上下移动时的摩擦引起的，类似于地球海洋潮汐中水的晃动。由于木卫二和木卫三的周期性引力推动，木卫一的轨道被挤压成椭圆形。木卫三、木卫二和木卫一绕木星一周所需的时间比为4∶2∶1，这导致这些卫星周期性地排成一列，并在引力作用下相互推动。这种关系被称为**拉普拉斯共振**，以法国数学家皮埃尔-西蒙·拉普拉斯的姓命名。我们在第二章也提到过他。

与木卫一类似，木卫二的轨道也被挤压成了椭圆形，所以它也会受到潮汐加热的影响。由于木卫二离木星更远，所以它受影响的程度要比木卫一小。木卫二冰层表面的平均温度为−155℃，但冰下深处可以通过潮汐加热保持液态水。人们利用木卫二冰面的陨石坑密度推算，木卫二冰面年龄仅为2 000万年到1.8亿年，表明有液态水从冰下喷出改造了冰面。图片显示冰面上存在一些**混沌地形**，即冰层破裂、冰块移动形成的地形，这表明表层之下的冰可能发生了对流。木卫二的冰面布满了条纹和脊，包括一些由于冰面被撕裂以及冰被挤压形成的带。木卫二表面存在硫酸镁盐，可能来自内部海水。冰面上还有一些红色条纹，目前还不清楚它的物质成分。

1995年至2003年，NASA的"伽利略号"探测器环绕木星进行的磁场测量为证明木卫二上有冰下海洋提供了关键证据。与地球或木星不同，木卫二的内部不会产生自己的**内秉磁场**。但当木卫二穿过木星的巨大磁场时，其内部会产生感应电流。反过来，这些电流会产生一个微弱且不断变化的**感应磁场**。人们测量了木卫二感应磁场的强度，发现这种强度要求在木卫二表面200千米以下区域存在导电流体。目前最有可能的解释是，存在盐水海洋，且体积是地球海洋的两倍。木卫二表面最大的陨石坑表明，其内部海洋上方的冰层至少有25千米厚。然而，一些突出的地形可能是由下部几千米处的融水上涌然后重新冻结形成的。所以，木卫二上可能存在透镜状浅湖。

除了液态水之外，地外生命的可能性还取决于能量来源、环境界面以及SPONCH等生物构成元素的可获得性。如果木卫二是由类似于富含碳的陨石或彗星的物质构成的，那么它可以提供充足的生物构成元素。此外，水和海底岩石的放热反应可以提供能量与营养物质。岩石中的放射性同位素（如钾、铀和钍）衰变产生的**放射性产热**可能会引起海底火山活动，为化能自养生物提供二氧化碳和氢气。

木卫二海洋中可能存在的氧气也支持海底发生铁和氢参加的生物氧化。木卫二海洋中有两个规模较小的潜在氧气来源。被困在木星磁场中的带电粒子会撞击木卫二冰冷的表面，破坏水分子，从而释放出氧气。如果木卫二的冰层发生搅动，一些表面产生的氧气可能会被带到海洋中。另外一个氧气来源是放射性元素的辐射分解水分子。在假定木卫二上存在生命的情况下，木卫二的海洋生物量最多只相当于地球的十亿分之一到一千分之一。不过，一些乐观主义者甚至推测，木卫二海洋中可能有足够的氧气供动物一类的复杂生命生存！

对木卫三和木卫四上感应磁场的检测也表明，它们存在冰下海洋，但比木卫二上的更深，分别在表面以下大约200千米和300千米处。通过陨石坑计数法推算得到的木卫三平均表面年龄大约是5亿年，所以它的下部海洋最近可能没有喷涌到表面。由于木卫三是木星最大的卫星，它内部巨大的压力导致深部的水形成了高压相的冰，位于海洋之下。因此，木卫三缺少深部的岩石-水界面，可能不像木卫二那样适合生命存在。木星对木卫三较弱的潮汐加热也暗示，木卫三的海洋中生物量相对较小。

比较令人意外的是，有些人认为木卫四也有冰下海洋。由于木卫四缺少潮汐加热作用，木卫四冰下如果有液态水的话，一定是由辐射能加热的，并且其海水中可能有较高浓度的盐来降低水的冰点。人们通过陨石坑计数得出木卫四的表面年龄是40亿年，所以在木卫四表面发现海洋生命证据的希望可能很渺茫。此外，相比于木卫三，木卫四上生命可利用的能量更少，因此潜在生物量也会更小。

土星的冰卫星：土卫二和土卫六

土星和它的62颗天然卫星①到太阳的距离大约是木星到太阳距离的两倍。在太空时代之前发现的比较大的卫星有（在土星轨道上由内向外）：土卫一、土卫二、土卫三、土卫四、土卫五、土卫六、土卫七、土卫八和土卫九。土卫二是其中比较特殊的，存在活跃的地质活动。它是土星的第六大卫星，但它的直径只有500千米，差不多与英国相当。土卫二绕土星公转的周期是

---

① 截至2024年12月，已经发现了146颗土星的天然卫星。

土卫四的二分之一。因此,土卫四的周期性引力推动迫使土卫二的轨道变成椭圆形,而土星变化的引力让土卫二变形并产热。目前尚不完全清楚为何土卫二冰层下的热量集中在南极地区。在那里,水冰颗粒和气体从被称为**虎纹**的平行裂缝中喷射出来。这些裂缝长约130千米,宽约2千米,两侧有冰脊。土卫二的喷出物中含有甲烷、氨、有机化合物以及盐类。事实上,这些喷出物为土星环中的E环提供了物质,而土卫二就在这个环内围绕土星公转。

土卫二有一个岩石内核、一个冰壳,以及喷流区域下面的冰下海洋。有机分子、能量和液态水的完美组合意味着土卫二内部可能存在生命。这些潜在生命可能是化能自养型的,依靠由水-岩石反应产生的氢或由水辐解产生的氢和氧为生。如果那里真的有生命存在,那么喷出物中的甲烷或有机物可能是生物产生的。

作为土星最大的卫星,土卫六也是天体生物学研究的热点。它的大小与木卫三以及水星相似,是太阳系中唯一拥有浓密的大气层的卫星。事实上,土卫六表面的气压为1.5个标准大气压,比地球表面的气压高50%。土卫六的大气中含有95%的氮和5%的甲烷,提供了10℃的温室效应,但到达土星轨道的阳光强度是地球的1%,所以土卫六的表面非常冷,低至-179℃。

土卫六的大气层中有碳氢化合物形成的有机雾霾。在高层大气中,太阳光中的紫外线可以分解甲烷分子,然后经过一系列光化学反应,生成多种碳氢化合物,包括乙烷、乙炔、丙烷、苯以及含有多环芳烃的红棕色颗粒。这些颗粒被统称为**托林**(tholins),来自希腊语tholos,意思是"浑浊不清的"。

这些甲烷的光化学产物最终从大气中沉降下来。事实上，土卫六大约20%的表面是类似于热带沙丘的地貌。这些"沙丘"是由有机物组成的，或由至少覆盖着有机物的沙子大小的颗粒组成的。在土卫六的两极地区分布着400多个湖泊，里面是液态丙烷、乙烷和甲烷的混合物，还有由苯和乙炔颗粒组成的特大海滩。

土卫六上的甲烷就像地球上的水一样，会形成云、雨和河流。我们对土卫六的大部分认识来源于2004年抵达的"卡西尼-惠更斯"计划中的土星探测器。其中，"惠更斯号"于2005年登陆土卫六，而"卡西尼号"是土星轨道飞行器。在"惠更斯号"着陆点附近，人们可以观察到液态碳氢化合物侵蚀形成的河道（图10）。在着陆点，（可能由水冰或干冰构成的）鹅卵石已经被磨圆，似乎经历了液态甲烷河流的搬运。土卫六表面罕见甲烷雨，但一下雨就是倾盆大雨。

由于阳光会持续破坏土卫六大气中的甲烷，如果没有有效补充，大气中的甲烷将在大约3000万年至1亿年的时间内耗尽。土卫六大气中的甲烷是如何得到补充的，这仍然是一个谜。目前流行的一个假说是，甲烷可以从土卫六内部泄漏出来，这暗示了土卫六内部存在活跃的地质运动。

土卫六太容易发生形变，因此不可能完全是固体的。这证明土卫六有一个地下海洋。土卫六轨道的偏心率使得土卫六受到土星引力的挤压。海洋存在于厚度不到100千米的冰壳之下。但和木卫三一样，土卫六的海底应该是致密的冰而不是岩石，因为土卫六的质量很大，足以使海底的冰呈高压相。

土卫六上可能存在两种类型的生命：冰下海洋中的类地生命和碳氢化合物湖中的怪异生命。关于第一种可能性：当土卫

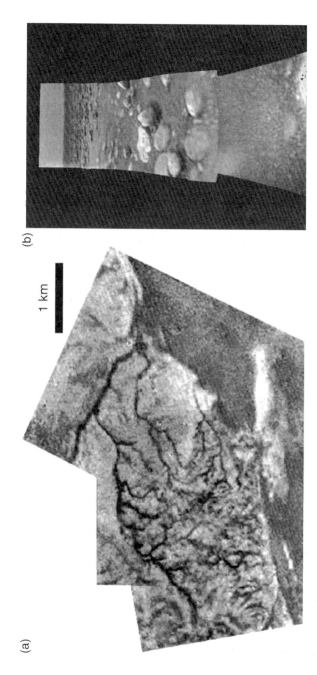

图10 (a) 一个河谷网络，似乎流向"惠更斯号"着陆点附近的一片平原。(b) "惠更斯号"着陆点的地表影像。前景的石头有10到15厘米大小，位于一片更暗更细粒的基质上

六形成时，它应该是通过捕获较小天体释放的热量而在其表面暂时维持了液态水环境。也许在土卫六的海洋逐渐隐入冰下的过程中，以水为基础的生命出现并存活下来了。

怪异生命应该难以具有足够的生物化学复杂性，因为土卫六表面缺乏氧原子，而氧恰恰是形成糖、氨基酸和核苷酸所必需的。当然，土卫六也可以从太空中捕获一些含氧分子，但这种补充量很小。这种假想的生命生活在冰冷的碳氢化合物溶剂中，依靠乙炔获得能量。在地球上，乙炔在氧气中燃烧是焊枪的工作原理。与在地球上不同，土卫六上的生物可能可以使用土卫六空气中的氢来代谢乙炔。化学反应如下：

$$C_2H_2（乙炔）+3H_2（氢）=2CH_4（甲烷）$$

这个反应可以产生能量，但这只是一种推测。我认为，土卫六上怪异生命面临的主要挑战是，大分子有机物在液态碳氢化合物中的溶解度很低，而大分子恰恰是形成基因组所必需的。在液态碳氢化合物中，大分子有机物通常比小分子更难溶。此外，在土卫六较低的温度下，有机物的溶解度也会进一步下降。

一些系外行星上可能有液态碳氢化合物，类似土卫六上的。其中，最常见的恒星是红矮星，它们比太阳小，温度也比太阳低。在这类矮星周围近1个天文单位处，系外行星的表面温度应该很低，在维持液态碳氢化合物，而不是液态水稳定的范围内。因此，如果真的在土卫六上发现了一些怪异生命，我们可以想象一个充满了与我们完全不同类型的生命的宇宙。讽刺的是，这样我们反倒成了怪异生命。

海卫一：围绕海王星运转的被捕获的柯伊伯带天体

在距太阳30个天文单位的海王星轨道之上及之外，天体往往被氮冰所覆盖。海王星13颗卫星[①]中最大的海卫一，以及冥王星都是**柯伊伯带天体**（KBOs）。**柯伊伯带**是黄道面内距太阳30至50个天文单位的冰冻天体分布区。这些天体都是太阳系形成时遗留下来的。目前，在距太阳1 000个天文单位的地方也发现了散落的柯伯伊带天体。考虑到它的两个轨道特征，人们认为海卫一是被海王星捕获的。首先，海卫一的公转方向与海王星的自转方向相反；其次，海卫一的轨道平面相对于海王星的赤道平面倾斜了157°。

由于海卫一反射了85%的入射太阳光，它的表面极为寒冷，大约为-235℃。它表面极其稀薄的氮气产生了十万分之二个标准大气压的大气压强。在这种温度下，所谓的"空气"其实是氮冰上方的氮蒸气。海卫一表面还有一些甲烷，大约是氮气浓度的0.03%。这些甲烷被太阳光中的紫外线破坏后，形成了一层薄薄的碳氢化合物颗粒雾霾。如果海卫一上甲烷的破坏反应以现今的速度持续了45亿年的话，那么这个过程大约会积累起来1米深的有机物质。然而，活动的霜冻可能掩盖了这些有机产物。不过，人们也在海卫一表面发现了一些红色的冰，可能代表光化学反应生成的有机物。事实上，氮冰被太阳光加热，可能会产生间歇泉般喷射的羽流，携带着可能是有机物的黑色颗粒。除了氮冰，海卫一表面也分布着水冰和干冰。

[①] 截至2024年12月，已经发现了16颗海王星的天然卫星。

海卫一存在活跃的地质活动，因为陨石坑计数表明海卫一的表面年龄只有1 000万年。此外，海卫一表面还分布着像喷口、裂缝和熔岩这样的冰结构。我们称这种火山结构为**冰火山**，意思是一种以融冰为岩浆的火山活动形式。对海卫一的密度探测表明，它有一个巨大的岩石内核，为冰火山活动提供了足够的放射性热量，并可能在冰下形成一个氨水海洋。

早期的海卫一可能更宜居。在刚被捕获时，海卫一的轨道可能是高椭圆的。此时，海王星对海卫一的引力可以产生巨大的潮汐加热作用，应该会融化出一个巨大的冰下海洋。随着时间的推移，潮汐作用会倾向于使海卫一的轨道变圆，而如今海卫一的轨道也的确接近圆形。因此，在捕获之后，随着潮汐加热作用的减弱，海卫一的冰壳逐渐变厚。现今，在海卫一数百千米厚的冰下可能仍然有液态水海洋，而海洋下方是高压相的冰。尽管海卫一上潜在的海洋生命可能深藏不露，但冰火山的存在表明，如果我们可以对海卫一表面的有机物进行采样，就有可能发现潜在生命的痕迹。

冥王星上有冰下海洋吗？

冥王星以冥界之神的名字命名，其表面主要是氮冰，大气主要是稀薄的氮气，压强为0.8到1.5千帕。就像在海卫一和土卫六上一样，太阳光也会破坏冥王星大气中的一些甲烷，并且产生碳氢化合物颗粒。这些光化学产物可以沉降到冥王星表面，这可能解释了其表面为何分布着一些黑色区域。

一般认为冥王星和另一个柯伯伊带天体之间的碰撞创造了冥卫一，也就是冥王星五颗卫星中最大的一颗。这次撞击事件有些类似于形成月球的那次撞击。因此，冥王星实际上与冥卫一构成了柯伊伯带中的一个双星系统：冥卫一的大小是冥王星的一半，质量是冥王星的九分之一。每隔6.4天，冥王星和冥卫一就围绕它们之间的一个点旋转一周。在冥卫一形成后，潮汐加热作用应该已通过融化使冥王星形成了冰下海洋。冥王星可能有一个岩石内核，可以提供足够的辐射能，来维持如今一个冰下富含氨的海洋。

冥王星海洋中的潜在生物会受到能量供给的限制，所以整体生物量要比木卫二上少得多。冥王星的海洋可能处在厚度超过350千米的冰下，很难在未来的太空任务中被直接观察到，但那里依然可能存在（过）生命。

我们对冥王星的研究表明，生命可能存在于极不可能存在的地方，因此我在表1中的总结可能还是太保守了。人们从未考虑过，许多寒冷的太阳系外天体是潜在的生命栖息地。但人们的观念已经发生了转变。目前，人们认为土卫五、天王星最大的两颗卫星天卫三和天卫四，以及一些大的柯伊伯带天体，比如塞德娜上都可能存在富含氨的冰下海洋。塞德娜的大小与冥王星相似，但距离太阳有100个天文单位那么远。鉴于我们一个太阳系中就有如此多的天体有生命宜居的环境，在我们的银河系中肯定存在数十亿个类似的可能宜居天体。

第七章

偏僻的世界，遥远的恒星

搜寻系外行星

在太阳系之外，天文学家已经发现了超过3 400颗系外行星，包括可能的和已确认的目标。[①]考虑到大型天体更容易被探测到，所以已发现的系外行星几乎都比地球大，有些还与太阳系行星截然不同。其中，**热木星**是距其母星0.5个天文单位以内的木星大小的系外行星，有的只需几天即可公转一周。还有一些系外行星听起来像大号的地球，叫**超级地球**，质量高达地球的十倍。虽然与地球大小相近的系外行星更难找到，但很明显，找到这类天体只是时间问题。我们不禁好奇，这其中有多少是宜居甚至存在生命的？

当然，在识别**宜居**的系外行星之前，你必须先找到系外行星。距离这么遥远，搜索起来十分困难，但天文学家已经找到了

[①] 截至2024年12月，已经发现了超过5 600颗已确认的系外行星。

两类方法。第一类**间接探测**是观察已知恒星的属性变化,例如位置或亮度的变化。这些属性会受到看不见的行星的影响,因而可用于间接确定未知行星的存在。第二类**直接探测**是利用图像或光谱直接观测行星的光。

四种间接探测方法是:天体测量法、恒星多普勒频移法、凌日法和微引力透镜法。前两种方法的关键是,有一颗行星和它的母星绕着共同的质心运行。比方说,如果一颗行星和母星的质量完全相同,那么它们将围绕它们连线的中点运行。实际上,行星往往要比母星小,因此质心更靠近母星,甚至可能在母星内部。但无论如何,这颗母星都会绕着质心沿一个小轨道运行。即使我们看不见它周围的行星,我们也可以观察到母星的"摆动"。对于我们的太阳系来说,木星每12年公转一周,会导致太阳位置发生以12年为周期的摆动。土星的公转给太阳叠加了另一个较小的、以29年为周期的摆动,对应它29年的公转周期。

天体测量法使用望远镜测量天空中恒星的移动。这种方法对距离母星较远的大型行星很灵敏,因此可以用来寻找与我们太阳系相似的星系。但是你必须等待几年或几十年才能追踪到这些行星的影响。

第二种技术,**多普勒频移法**或**径向速度法**,主要原理是当恒星的光分成各种颜色时,它的光谱会像条形码一样具有暗带。这是由于恒星大气中的元素会吸收来自恒星内部的特定波长的光子,从而产生一些暗线。如果恒星正在靠近或远离地球,这些线将分别移动到更高(更蓝)或更低(更红)的频率,这被称为**多普勒效应**。其实,我们每个人都体验过声音的多普勒效应。当路上一辆鸣响警笛的警车靠近你时,声波会集中成高频的尖啸声,但在警车经过后,又变成了低频的嗡嗡声。光也会发生类似的频移。恒星光谱暗线有节奏的红移和蓝移表明,这颗恒星正在摆动并且周围有一颗行星。谱线位移的大小可以指示行星的质量,而其速率表明了行星公转一周的时间。

1995年,来自日内瓦大学的迪迪埃·奎洛兹(当时还是学生)和他的导师米歇尔·梅厄利用多普勒频移法探测到了第一颗围绕类太阳恒星运行的系外行星。这颗行星的质量至少为木星的一半,每四天围绕飞马座的一颗恒星运行一周。这是一项极为惊人的发现。没有人能想到一颗巨行星可以离它的母星这么近(0.05个天文单位),因为巨行星不应该在这么近的距离形成。现在我们了解到,一些系外行星在其演化早期经历了行星迁移(详见第三章),而我们最终看到的是轨道来回推挤后稳定下来的轨道与幸存的系外行星。2012年,多普勒频移数据表明,可能有一颗质量与地球相当的行星,距离我们只有4.3光年,在绕着半人马座α星B这颗恒星运行。如果这颗系外行星被证实存在的话,它距离半人马座α星B只有0.04个天文单位。距离如此之近,它的表面很可能是熔岩。①

多普勒频移技术的一个微妙之处在于我们观察行星系统的方式。如果一个行星的轨道是正对我们的(也就是说,轨道面相对于垂直于视线的平面的倾角为零),我们就看不到摇摆往复的多普勒频移。轨道越侧对我们,我们可以观察到的多普勒频移越大。当上述倾角达到90°时,我们观察到的多普勒频移最大。通常我们

---

① 作者将此处更新为:"2016年,多普勒频移数据表明,仅仅4.25光年外就存在一颗与地球质量相当的行星,它围绕我们最近的恒星比邻星公转。这颗行星被命名为比邻星b,它距其恒星0.05个天文单位,正处于恒星的宜居带,即理论上行星表面可以维持稳定液态水存在的区域之内。更多数据表明比邻星可能还有另外两颗行星,但是还未确认。"

无法知道倾角的大小,因此测得的多普勒频移可能小于理想的侧对情况。所以,通过这种方法,我们只能推断出系外行星质量的下限。这种多普勒技术对靠近母星的大型行星最灵敏,所以最初发现的大多数系外行星都是热木星。但是,我们现在通过凌日法发现,热木星实际上只是系外行星中的极少数,大约为0.5%到1%。

如果我们足够幸运,是从几乎完全侧着的方向观察一个系外行星系统,那么就可以使用**凌日法**,测量行星掠过恒星时恒星亮度的下降值。从统计学上看,这种位置布局是罕见的,但宇宙中有这么多恒星,如果你观察的范围足够广并且时间足够久,就可以看到凌日现象。如果母星的大小能估算出来,那我们可以通过亮度降低的数值来确定凌日的这颗系外行星的直径。接着,这颗行星的轨道周期和它与母星的距离也可以利用光变暗的周期计算出来。NASA的"开普勒"计划是一台围绕太阳运行的望远镜。自2009年到2013年,它一直凝视着天鹅座中的145 000颗主序星以寻找凌日现象,它们大多在500到3 000光年之外。开普勒天文望远镜已经发现了3 200多个系外行星候选者。[①]这些候选者可以通过验证周期性变暗或使用其他方法(例如多普勒频移法)来确认。2017年,凌日系外行星巡天卫星(TESS)望远镜[②]将开始检验200万颗恒星的凌日情况。[③]

第四种间接方法是**微引力透镜法**。爱因斯坦的广义相对论预测,当遥远的恒星和我们之间有一个物体经过时,该物体的引

[①] 开普勒天文望远镜于2018年11月退役。截至当时,共发现5 011个系外行星候选者、2 662个已确定的系外行星。
[②] 实际上于2018年4月发射。
[③] 作者在此处新增的内容为:"2021年,韦布空间望远镜成功发射,原则上它可以探测系外行星大气的信息,并且测量红矮星宜居带内凌日行星更详细的大气成分。"

力场会使光线发生弯曲。这个前景物体有效地起到了透镜的作用,即聚焦光线并使远处的恒星看起来逐渐变亮。但如果透镜星有一颗环绕它的行星,那么背景星就会不止一次地变亮。这种灵敏的技术可以用于在距离恒星1.5到4个天文单位的轨道上找到与地球质量相当的系外行星。遗憾的是,它们一旦对齐,我们就几乎不可能再跟踪进行后续的详细测量,因为这些天体通常太遥远了。不过,使用微引力透镜法还是能得到一些统计信息的。在星际空间中独自漂泊的大型行星也可以作为透镜物体,由此带来的一个发现是,似乎有许多不受约束的、接近木星质量的行星在不同恒星系之间飘荡。这些孤独的世界大概是从它们原来的恒星系中被弹射出去的。

对于天体生物学研究来说,最重要的技术是**直接探测**,即捕获从行星来的光。但直接探测面临很大的技术挑战,因为行星是暗淡天体,而一旁的恒星则明亮得多。尽管如此,还是有一些太空望远镜,例如哈勃望远镜,以及地面上的超大型天文望远镜实现了直接探测。它们使用了一种叫**日冕仪**的设备来遮挡恒星的光。这种设备最早被用于遮挡太阳光以研究日冕,即我们在日食中看到的微弱太阳光晕。地面望远镜还使用**自适应光学**技术,可以探测并纠正由地球大气波动引起的畸变。

天基望远镜的优点是,可以避免地球大气层导致的图像模糊,但问题是,我们应该研究光谱的哪一部分。一般来说,行星主要发射红外辐射并反射可见的星光。如果我们从遥远的太空看我们的太阳系,会发现太阳又热又大,在红外光波段比地球亮约一千万倍,在可见光波段比地球亮约一百亿倍。因此,在红外光波段中寻找遥远的地球信号要比在可见光波段中对比度高一千倍。遗憾的是，当光线遇到任何物体（例如望远镜），都会扩散和变模糊（衍射），而这种效应在红外光波段比在可见光波段更明显。幸运的是，一种被称为"干涉测量"的技术可以消除不需要的光。我们用的降噪耳机也依据相同的原理来消除不需要的声波。

光波像水波一样有波峰和波谷。**归零干涉测量法**使用多个望远镜镜面来精确调整从天空中同一个点传来的光波，从而使波峰波谷相消，让图像的光亮处变暗。通过这种方式，恒星的光可以被"归零"，我们从而可以观察到系外行星。NASA 和 ESA 都研究了应用干涉测量法的太空望远镜——分别被称为**类地行星发现者**和**达尔文**①，计划在太阳系附近捕捉类太阳恒星周边系外行星的图像和光谱。

迄今为止，人们对于系外行星的调查结果普遍表明，低质量的系外行星数量在增多，因此地球大小的岩质行星在宇宙中应该比较常见。对于系外行星，我们可以通过密度来推测其内部物质组成。前人通过多普勒频移法获得系外行星的质量，结合通过凌日法获得的体积，可以进而推算它的密度。此外，如果一颗地球大小的系外行星能够发生凌日，并且至少有一个类似海王星的兄弟姐妹，那么我们还可以利用一种被称为**凌日时间变分**的方法来估算该系外行星的质量，从而得到密度。即使较大的系外行星由于其轨道倾角而没有从其母星前掠过，它也可能导致同星系的较小行星的凌日时间略微发生变化。我们也可以由此估算出系外行星的质量和密度。这些对系外行星密度的测量，会为我们理解系外行星是否由气体、岩石、水，或者这些物质

---

① 这两项天文探测计划均已被取消。

的混合物构成提供更多线索。

宜居带

当我们考虑太阳系中的生命宜居性时，我们专注于含有液态水的天体，因为我们确信这是形成类似地球生命的必要条件。同样的思路也适用于寻找宜居的系外行星。我们的太阳系内有一些表面干燥但存在地下海洋的行星，很值得用空间探测器进行天体生物学研究，但目前有关系外行星的天体生物学通常不会以这类天体作为首要目标。因为，在不太遥远的未来，我们都没有希望可以发射飞行器造访系外行星，能做的只有用望远镜观察远处系外行星发出的光。因此，我们现在最关心的是地表有液态水的系外行星。这些行星有可能支持一个生物圈，它会向大气中释放一些生物成因的气体。原则上说，这些由生物产生的气体可以通过分析系外行星的光谱来检测到，从而指示生命的存在。

1853年，威廉·休厄尔指出，地球与太阳的距离介于他所谓的"中央热区"和"外部冷区"之间，故液态水可以稳定存在。在20世纪50年代，美国天文学家哈洛·沙普利（1885—1972；他首次正确估算了银河系的大小及太阳所处的位置）也谈到了恒星周围存在一个地带，处于这个地带的行星具有利于液态水稳定存在的温度条件。如果一颗行星离它的母星太远，它表面的温度就会太低，导致液态水结冰；如果距离太近，它表面就太热了，无法存在稳定的液态水。如今，**宜居带**一词是指在某个时刻，恒星周围可以支持类地行星表面存在液态水的环带区域。我们之所以指定一个特定的时刻，是因为恒星会变老并且变亮或变暗，因此它的宜居带会移动。不同于宜居带的定义，**连续宜**

居带指的是在某一段时间内，通常是恒星的主序阶段，恒星周围可以支持行星宜居的环带区域。

科学家已经估计了包括太阳在内的不同类型的主序星周围的宜居带宽度。总体来说，恒星宜居带的内边界的位置取决于行星在此处会发生失控的温室效应，而外边界的位置则取决于行星表面无法通过温室效应有效地保温，例如，大气中的二氧化碳在低温下凝华成了干冰云，或大气中的二氧化碳浓度太高以至于反射了大部分的入射阳光。如我们在第六章中讲到的，第二种温室效应失效的情况是早期火星面临的一个潜在问题。对于我们的太阳系来说，宜居带的内边界约在距太阳0.85到0.97个天文单位的位置，而外边界为1.4到1.7个天文单位。二者都有不确定的范围，分别反映了水和二氧化碳云对内边界和外边界位置的影响。例如，云可能会通过反射阳光来冷却靠近太阳的行星表面，因此内边界可能在0.85个而不是0.97个天文单位的位置。目前最乐观的估计把火星归入了宜居带的外边界以内。然而，火星较小的体积导致它的大气层很稀薄，无法维持表面存在稳定的液态水。换句话说，如果火星和地球一样大，它现在可能是一个宜居星球。所以行星大小对于行星的宜居性演化也很重要，而是否处于宜居带内并不是行星宜居性的唯一参考，而只是初步指南。

在其他恒星周围，宜居带与恒星的远近取决于恒星的温度。例如，温度比太阳低的K型恒星（图1）有一条很窄的宜居带，相当于太阳系中水星轨道到地球轨道的距离。一个温度更低的M型红矮星的宜居带中心距该恒星只有0.1个天文单位左右，比太阳系中0.4个天文单位处的水星轨道还近得多。

事实上，这些暗淡的M型矮星的宜居带太贴近它们了，以至于宜居带中行星的自转周期会受矮星引力作用的强烈影响,行星因而发生**潮汐锁定**。这种现象也发生在地球-月球系统中,月球每绕地球公转一圈,就会自转一圈,导致月球总是同一面朝向地球。这种现象被称为**同步自转**。月球具有相同的公转周期和自转周期并非巧合。月球在地月连线方向上稍远一些,如果它的长轴稍稍偏离地月连线,地球的引力会将其偏离的部分扭回来,重新对齐。

如果一颗处于宜居带中的系外行星与M型矮星发生了同步自转,这颗行星的一面会一直对着阳光,而另一面将永久处于黑暗中。然而,这并不意味着这样的行星不宜居。虽然处于黑暗中的半球会很冷,但只需要厚度适中的大气层来输送足够的热量加热夜半球,同时云层可以帮助冷却昼半球。此外,如果这个行星有一颗大卫星绕其公转,这也许可以影响阳光的分布。因此,对于生命宜居性来说,M型矮星宜居带内的潮汐锁定现象并不是致命性问题。这个发现鼓舞人心,因为M型矮星是宇宙中最常见的一类恒星,约占恒星总数的四分之三。

近些年来,人们开始质疑一些用于估算宜居带范围的假设条件。通常,我们将宜居带的外边界设定为二氧化碳开始发生凝华的位置。但是一些大型岩质系外行星可能有厚厚的富氢大气。我们通过研究土卫六和巨行星发现氢和甲烷也是温室气体,并且它们的凝华温度比二氧化碳要低得多。此外,对于宜居带内边界的传统定义也存在问题。例如,一个只有极地湖泊而非全球性海洋的贫水行星可能不会发生失控的温室效应,因为它没有足够的水来产生浓密的水蒸气大气,从而无法完全阻挡红外辐射。因此,宜居带可能比我们想象的更宽。

是否存在银河系宜居带?

除了恒星宜居带,一些天文学家还提出,银河系中有一个最适合宜居行星系统的区域,称为银河系宜居带。他们注意到,太阳系处于距银河系中心三分之二个银河系半径的位置,可能有利于形成宜居的行星系统。如果一个行星系更靠近恒星密布的银河系中心,它的宜居性会受到超新星或过往恒星的干扰。另外一种极端情况是,如果一个恒星系靠近银河系边缘,它将会非常缺乏除氢和氦之外的元素,不利于行星的形成。这里不得不说一个天文学家的奇怪习惯(它让化学家十分头疼)。他们会将氢和氦以外的**所有**元素称为"金属",并将恒星中这些元素的含量称为**金属丰度**。如果遵循这个说法的话,拥有巨行星的恒星系往往有丰富的金属。人类首先发现的系外行星都是巨行星,因而人们最初认为星云中行星的形成需要高的金属丰度。不过,最近的研究发现,与太阳系质量相似的恒星系是否拥有超级地球等大型系外行星,和它的金属丰度并没有必然联系。

银河系宜居带的概念还存在另外一个问题,那就是恒星不会原地不动。近期,有研究表明,银河系旋臂的引力散射会导致恒星在星系盘上徘徊游荡。无论如何,在可预见的未来,人类只能深入研究那些靠近太阳系的系外行星来寻找地外生命信号,比如距离不到100光年的。因此,无论银河系宜居带是否存在,它都不是短期内寻找宜居系外行星时应该参考的实际因素。

生命信号,或者如何寻找存在生命的行星?

诚然,系外行星研究的核心天体生物学问题是我们能否发现地外生命。1990年，NASA的"旅行者1号"探测器在朝太阳系之外的方向飞行时，从距地球轨道61亿千米处回望了地球，并拍摄了一张著名的照片《暗淡蓝点》。这张照片上，地球小到只是一个蓝色像素。如果我们并不知道这个星球的其他信息，仅凭这张照片，我们能推断出什么？如果我们说这颗未知行星的颜色反映的是蓝色的海洋和白色的云层，那仅仅是我们的猜测而已。我们怎样才能了解得更多？

一般来说，我们需要一颗行星的可见光或红外光谱来确定它的特征并寻找潜在生命。具体原理是，不同的分子和原子可以吸收光谱中不同频率的光。据此，我们可以通过分析光谱来寻找大气中那些由生物产生的气体，例如氧气或甲烷。这些可以指示潜在地外生命活动的行星特征被统称为**生命信号**。事实上，在"旅行者1号"拍摄《暗淡蓝点》的同一年，前往木星的"伽利略号"就采集了地球的光谱数据。这是一种为系外行星探测所进行的演练，证明人类可以在遥远的太空探测到地球大气中的氧气、甲烷和丰富的水。大气中同时有氧气和甲烷是存在生命活动的证据，因为这两种气体会迅速发生反应，从而消耗完其中一种气体，除非有生物活动在大量产生和供应它们。因此，我们说地球大气处于**化学不平衡**状态。这种不平衡的大气特征就是我们想要在系外行星探测时观察到的那种生命信号。

通过分析一些凌日系外行星从其母星后面和前面经过时的光谱差异，科学家已经收集了一些关于系外行星大气的信息。当行星从其母星前面经过时，即**凌星主食**发生时，人们收集到的光谱包含了那些穿过该系外行星大气环的光信号。主食结束后，采集恒星本身的光谱数据，再将主食时采集的光谱减去恒星本身的光谱,就可以分离出来系外行星大气的吸收光谱。另外一种方法是,在行星位于母星旁边时采集两者共同的光谱,然后减去行星位于母星后方,即**凌星次食**发生时测得的光谱,也可以得到系外行星整体的光谱。事实上,这项技术已被应用在围绕太阳运行的斯皮策太空望远镜上,用于获得热木星的红外光谱。

借助更先进的望远镜,我们希望可以确定类地系外行星的表面温度以及是否有液态水。这些天体的表面光谱也许能给出这些信息,但在通常情况下,你也许只能看到来自系外行星云顶或高层大气的光。事实上,金星表面就是这样被厚厚的大气包裹着。对于同样被大气遮盖的系外行星,我们需要从光谱中的吸收线推断大气中温室气体的浓度,并利用这个浓度来计算表面温度。我们进一步结合系外行星的大气成分,可以推断表面是否存在液态水。我们也许还可以通过闪光信号来寻找海洋,闪光就是在掠射角度下,平滑的水体表面的明亮反射点。最后,我们应该留意一些**反生命信号**。微生物很容易代谢掉氢气或一氧化碳气体,因此二者中任意一种气体的富集都应被视作反生命信号,可以用以排除该行星上存在微生物的可能性,即使它在宜居带内。基本上,反生命信号表明"这里没有生命"。

"搜寻地外文明"计划

人类寻找系外行星上地外生命的另外一个思路是假设存在技术文明。如果他们真的存在,我们可以寻找他们的通信信号。这些努力被统称为"搜寻地外文明"计划,或更笼统地说是搜寻**技术信号**的计划。

1959年,朱塞佩·科科尼和菲利普·莫里森主张使用大型射电望远镜寻找地外文明的广播信号。此后,也有人建议寻找潜在的使用可见光传输的地外文明通信信号。出于一些实际考量,目前人们只是在这些波段寻找通信信号。例如,X射线在高层大气中会被完全吸收,而无线电和可见光容易产生,也容易穿透大气从而被探测到。"搜寻地外文明"计划面临的主要挑战是地外文明是否值得寻找。宇宙中有足够多的文明吗?

1961年,彼时在康奈尔大学任教的弗兰克·德雷克提出了一种用来评估银河系中射电文明的潜在数量的方法。如果我们可以估算每年"开播"的文明的平均数量及这些文明的平均寿命,通过将这两个数量相乘,就可以得出当前射电文明的数量。打个比方,如果一所大学每年的新生人数在6 000人左右,他们平均在校园里待四年,那么任何时候这所校园里的本科生总数都是6 000×4=24 000。德雷克进一步精细化了计算,设计了六个因素来确定每年出现的射电文明"新生"的数量。总而言之,我们通过将六个数字乘以平均寿命来计算N,即通信文明的数量,如下所示:

$$N(文明数) = \underbrace{R_* \times f_{行星} \times n_{宜居} \times f_{生命} \times f_{智慧} \times f_{文明}}_{每年出现的通信文明的数量} \times \underbrace{l}_{通信文明的平均寿命}$$

这就是著名的**德雷克方程**。第一个数字R_*是适合承载生命的恒星的诞生率。天文学家每年观测到大约十颗新的G型、K型和M型恒星。其他因子如下:

$f_{行星}$是上述恒星中拥有行星者的比例;

$n_{宜居}$ 是每个行星系统中宜居行星的平均数量;

$f_{生命}$ 是上述行星中可以诞生生命并发生演化者的比例;

$f_{智慧}$ 是存在生命的行星中能出现智慧生命者的比例;

$f_{文明}$ 是上述出现智慧生命的行星中能发展出星际通信文明者的比例;

l 是这些通信文明的平均寿命。

等式中有几个因子是未知的。但是,管它呢!我们还是先试试看这个方程。目前系外行星探测结果表明,至少三分之二的恒星都有行星,所以我们将 $f_{行星}$ 设定为2/3。人们还在分析"开普勒"计划观测到的数据,但已有的结果表明每100颗行星中至少有1颗是宜居的,所以我们把 $n_{宜居}$ 设定为1/100。现在,我们只需要猜测其他几个参数。地球上的生命很快就诞生并演化了,所以我们假定有一半的宜居行星可以孕育生命,即 $f_{生命}$=1/2。我们假设,所有生物圈中能够演化出智慧生命的比例是 $f_{智慧}$=1/8。我们再猜测一下,假定十分之一的智慧生物圈可以发展出有能力进行星际传输的文明,所以 $f_{文明}$ 是1/10。最后,通信文明的平均寿命 l 需要社会学层面的假设,不过让我们先假设这个数是10 000年。那么我们在银河系中可以有多少个通信文明呢?答案是10×(2/3)×(1/100)×(1/2)×(1/8)×(1/10)×10 000≈4。还不错。但是我们能不能更好地约束一下这些参数的不确定性呢?

对系外行星的研究将最终限定德雷克方程中的第二项和第三项参数,并且如果检测到生命信号,也许能为限定地外生命出现的概率 $f_{生命}$ 提供一些线索。乐观主义者们认为,一个拥有液态水和合适的物质的行星很容易孕育出生命。不过,目前我们实在不能确定这一点。

下一项是关于智慧生命的,这个概率是多大呢?一个相关的观察可能是,要想满足一些特定的生态需求,地球生物只有少数演化方案。在动物学中,我们经常发现一些不相关谱系的生物却有着相同的性状,这就是所谓的**趋同演化**的结果。趋同发生在特定的生物功能和生态位上。就视觉而言,至少有40个不同的动物群演化出了眼睛。狗就是一个趋同生态位的例子。与墨西哥狼(一种胎生哺乳动物)相似,袋狼(一种有袋动物)演化出了类似于狗的特征。演化的趋同性是如此普遍,意味着拥有腿、一对眼睛并且适应特定生态环境的生物体的出现也许是必然的,因为只有这样躯体才能解决行走、立体视觉和占据特定生态位的问题。

目前关于地球上技术文明的出现是独一无二的还是趋同演化的结果还存在争议。在地球历史上,它出现得很晚。地球花了40亿年的时间才积累了足够的大气氧气来支持动物生命,然后又过了几亿年才有了技术生命。其间,恐龙大约统治了地球1.7亿年,但我们没有发现技术出现的迹象,既没有化石工具,也没有恐龙制造的微波炉。不过,有一些谱系的恐龙似乎已经有了价值比较大的、发育了的大脑。但单纯的大脑质量不是衡量智力的良好标准,因为巨大的身体往往需要比较大的大脑来控制它。然而,我们可以用脑化指数来衡量智力。脑化指数即一种动物的脑质量与计算出来的该动物体重对应的参考脑质量的比值,这里的参考脑质量可以通过一些参考动物的体重和脑质量的相关关系预测得到(图11)。因此,一个普通动物的脑化指数为1。该数值高于1的动物为有头脑的,低于1的为没头脑的

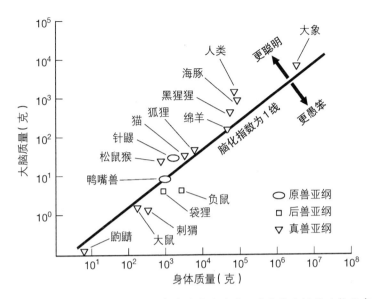

图 11 不同哺乳动物的大脑与身体质量的关系。对角线左侧的动物具有平均水平以上的脑化指数

（图 11）。人类和黑猩猩的脑化指数分别约为 7.4 和 2.5，这意味着他们的大脑比按拟合曲线估算的参考值要大。兔子的脑化指数为 0.4，爱发先生[①]要是知道这个可得乐坏了。动物中，正如亚里士多德所指出的，"人的大脑与体型的比率是最大的"。

古生物学家发现某些谱系的脑化指数随演化历史的推移而增加，最明显的是**智人种**。例如，生活在大约 200 万年前的**能人**，其脑化指数只有 4。齿鲸（海豚、抹香鲸和虎鲸）的脑化指数在大约 3 500 万年前有一次提升，当时它们演化出了回声定位，用来寻找鱼类和同伴。"同伴"可能是这次演化的关键驱动力。一般来说，社会性动物的智力比较高，如海豚的脑化指数为 5.3。智力的提

---

① 《乐一通》系列中的卡通角色，是兔八哥的对头。

升可能有助于这些动物生存及求偶，这又能进一步促进它们繁衍更多的后代。然而，关于智力背后的演化压力仍然有很多争论。

最后是关于"搜寻地外文明"计划的另一个质疑，它被称为**费米悖论**，由曾在1942年建造出世界上第一个核反应堆的诺贝尔物理学奖得主恩里科·费米（1901—1954）提出。他的想法源于一次午餐时的讨论。银河系盘中的恒星可以追溯到90亿年前，而地球只有45亿年的历史。因此，如果智慧生命很普遍，许多技术文明应该远早于我们出现。假设他们发展出了太空旅行技术或发明出了可以自我复制的机器人来代替他们做太空旅行，他们现在应该已经遍布整个银河系了。"他们在哪儿呢？"费米问道。

费米悖论有三个可能的解释。第一，由于德雷克方程中的一个因子小得惊人，文明在银河系中是个例而非普遍情况。第二，费米的前提不正确。由于各种原因，文明可能不会在银河系中四处殖民。第三，上述文明确实存在，但他们对我们或我们中的大多数隐藏了他们的存在。这其中，科学家们尤其不喜欢第三种解释，因为它是一种无法验证的强行自圆其说，但它受到了科幻作家、小报以及据调查三分之一美国成年人的喜爱。这些美国成年人相信外星人已经来过地球。但是，前两个才是更合理的解释。显然，所有为寻找地外生命所做的努力都与我们是否孤独，也就是第一种解释有关。如果"搜寻地外文明"计划成功，并且我们能够理解他们的通信信号，我们也可能深入了解其他文明。目前，已知的生命只存在于地球上，所以像费米一样，我们只能问："他们在哪儿呢？"

第八章
争议与展望

"稀有地球"假说

天体生物学中一些重大但悬而未决的问题往往会引发争议。其中一个例子出现在2000年，我在华盛顿大学西雅图分校的同事彼得·沃德和唐·布朗利出版了一本畅销书《稀有地球》，引发了一场争论。大体上，他们提出的**"稀有地球"假说**是说：支持地球上出现复杂生命的偶然环境是如此罕见，以至于地球可能是银河系中唯一存在智慧生命的星球。他们的论据包括：地球处在银河系中一个恰当的位置；太阳系中有木星去捕获可能会撞向地球的彗星；地球上板块构造所导致的独特的挥发分循环过程维持着大气层的存在；一系列偶然事件导致大气变得富氧；地球幸运地拥有一个巨大的月球，从而稳定了轴倾角，最终拥有稳定的气候。

《稀有地球》一书是对**哥白尼原则**的抨击。该哲学思想以尼古拉·哥白尼的名字命名，他提出的日心说推翻了地球处于宇宙中心这一前人的观点。哥白尼原则的拥护者认为，我们在宇宙中的位置没有什么特别之处。天文学家也提出了几点来支持这个原则。首先，地球肯定是宇宙中许许多多岩质行星里的一颗。此外，我们的太阳，一颗G型恒星在宇宙中也没有什么特殊的，因为宇宙中十分之一的恒星为G型。在银河系众多旋臂中，我们所处的位置也平淡无奇。正如斯蒂芬·霍金所言："人类只不过是一颗中等大小的星球上的化学浮渣，后者绕着千亿星系之一外围的一颗非常普通的恒星运转。"

哥白尼原则的支持者也提到，科学的巨大进步一步步彰显了假设我们是特殊的这一思想的愚蠢。我们只须想想伽利略的《星空信使》（1610）。伽利略在这本书里描述了一些对木星卫星和月球表面的观察，表明这些天体并不是神学家和哲学家们所谓的完美天体，反而可以用在地球上得到的物理规律来解释。同样，达尔文的《物种起源》（1859）也推翻了人类是独立于其他生物之外的存在的观点。

"稀有地球"假说的问题在于，它假定人们掌握了很多宜居性方面的知识，但事实上其中很多都是不确定的。比如说，最近人们发现，恒星的运动意味着，银河系的宜居带并不像之前认为的那样是一个很确定的概念。NASA的"开普勒"计划的探测结果也表明，除了类地行星之外，围绕恒星公转的其他类型的行星也很常见，这其中包括许多木星大小的天体。此外，在合适的位置存在一颗木星大小的行星，它可以降低彗星或小行星撞击地球的频率，这一观点最近也受到了挑战。众所周知，木星确实是像宇宙吸尘器一样，可以清除轨道附近的一些撞击体。例如，1994年，苏梅克-列维9号彗星撞向了木星，而在2009年和2010

年,人们在木星上看到了更多的撞击痕迹。目前比较清楚的是,木星的引力可以使彗星偏离原定轨道,并将其抛射出来;这些彗星来自一个由冰状物体组成的光晕,即在约50万个天文单位的距离上环绕太阳系的**奥尔特星云**。然而,对于地球来说,75%以上的撞击威胁来自太阳系平面内的近地小行星与彗星,木星的存在反而会使这些小天体的运行变得不稳定。因此对于地球宜居性来说,木星的作用是双刃剑。一些计算甚至表明,木星的净作用实际上是负面的,它并非我们的"朋友"。

此外,书中涉及的其他一些支持"稀有地球"假说的观点也是模棱两可的。在太阳系中,地球的确有独一无二的板块构造活动。但是,就板块构造的触发机制而言,一颗行星必须足够大,才能有充足的内部热能驱动板块运动。除此之外,板块运动可能还需要海水来冷却大洋板块并作为运动的润滑剂。在太阳系中,的确只有地球满足这些触发条件,但这并不意味着,系外宜居带中的类地行星不可能具有适合类似构造活动的条件。

虽然"稀有地球"假说准确地提出,没有大卫星的行星将比地球经历更大的轴倾角变化,但这种变化可能对行星上低纬度地区的气候影响较小。事实上,更厚的大气层、更广阔的海洋及更低的系外行星自转速率,可以消除行星两极和热带地区因轴倾角变化而产生的气候差异。最后,氧气没有在其他类地行星上积累起来这一点,可能会与"稀有地球"假说的假设条件背道而驰。有些系外行星可能比地球更适合演化出富氧大气,因为它们的火山活动喷出的可以与氧气反应的气体更少,因而大气中更容易积累氧气。

"稀有地球"假说可能正确的一点是,宇宙中类似微生物的生命应该比智慧生命更普遍。微生物的新陈代谢活动范围大得惊人，相比于复杂生命，它们能生存于更加广泛多样的环境中。不过，我们无论如何也应该对任何关于复杂生命普遍存在的确定性表述表示怀疑，这些表述都缺乏数据支持。正如卡尔·萨根的名言所说："保持开放的思想是有好处的，但不要脑洞大得让脑子掉出来了。"

展望未来：天体生物学研究与寻找地外生命

天体生物学研究令人兴奋的地方在于，它试图回答诸如生命起源以及我们在宇宙中是否孤独等问题。随着科技的进步，在未来几十年里，人类在该方向上取得重大发现的可能性越来越大。

在早期生命领域，人类将有可能在实验室中制造出现实可行的、能够自我复制的基因组。这将为我们了解生命起源提供深刻的见解。有几个团队正在研究"RNA世界"假说及其变体模式。此外，就南非和澳大利亚的古老沉积岩进行的深钻项目，必然会助力人类理解最早的生命及其所处环境。

在太阳系中，全球航天机构最优先考虑的目标之一应该是土卫二。土卫二有能量来源（潮汐加热）、有机物质和液态水。这是一份对生命所需条件的教科书式清单。除此之外，大自然还为天体生物学家提供了最好的"免费午餐"：将土卫二上的有机物质喷射到太空中的喷射流。以现有的技术，我们完全可以制造一台绕飞土卫二的航天器，并对其喷射流中的有机物进行采样分析。当然，如果这些物质可以被送回地球进行分析那就更好了。

事实上，发射航天器原位收集地外样本并将样本送回地球，即执行**采样返回**任务，是研究火星与金星历史以及它们是否有过生命的未来方法。目前看来，金星的采样返回任务还遥遥无期，NASA 和 ESA 现行计划中的一项战略目标就是完成火星的采样返回任务。

在木星轨道上，木卫二可能是最有潜力发现生命的卫星。对未来的木星探测而言，可以首先发射木卫二轨道飞行器来详细研究这颗卫星，并确定冰下海洋或湖泊上方的冰层厚度。①接下来可能发射着陆器，可能使用通过辐射热量发生器来融化冰层的机器人。最终，人们可以设想潜水器一头扎入木卫二的海洋中。

除了土卫二，土卫六也将是土星卫星中天体生物学研究的目标。②如果人类未来可以向土卫六发射一台湖泊着陆器，一种类似星际飞船的航天器，让它漂浮在土卫六极区的湖泊上，从而确定这些有机液体的具体成分，那将是科学上的一个巨大飞跃。此外，未来的土卫六轨道飞行器可以探测并确定土卫六地表下海洋的深度，并进一步研究土卫六的表面。

关于未来，可以确定的一点是，对系外行星的探索将持续激发人们对天体生物学的兴趣。我预料，人类将在其他恒星的宜居带内发现许多类地行星、几乎纯二氧化碳大气层的死气沉沉的系外行星、完全被波光粼粼的海洋覆盖的水世界，以及一些处于演化早期的类金星行星，这些类金星行星正在发生失控的温

---

① 2023年和2024年，ESA和NASA分别发射了"果汁号"和"欧罗巴快船号"，将对木卫二进行遥感和飞掠探测。

② 作者在此处新增的内容为："事实上，NASA的'蜻蜓号'探测器将于2034年在土卫六近赤道地区着陆。'蜻蜓号'是一台带有四个双旋翼的无人飞行器，它将飞越土卫六表面的不同地区，测量有机物，包括可能与生命起源相关的各种分子的化学数据。"

室效应,海洋中的水不断蒸发进入太空。

当20世纪90年代天体生物学作为一门新学科而崭露头角时,有人曾质疑它的未来,并怀疑它会不会由于没有很快找到地外生命或无法回答有关生命起源的问题而最终昙花一现。然而,人类对于系外宜居带中地球大小的系外行星的发现让我们确信:宇宙中其他地方存在生命的可能性变得比以往任何时候都更有价值,更有意义,更与我们息息相关。天体生物学正是为此而来,也将继续前进。

本译作是中国科学技术大学天体生物学实验室的集体成果,也是同学们自发组织的"南七译站"的首部作品。在翻译的过程中,我们得到了陈至仪、周强、张成涛、雍骋正、刘士言等同学的无私帮助,也获得了黄方、林巍、应见喜、罗根明等老师的指导,在此表示由衷的感谢。

——郝记华　张晟霖

索引

（条目后的数字为原书页码，见本书边码）

A

acetylene on Titan 土卫六上的乙炔 106
adaptive optics 自适应光学 113
aeons on Mars 火星的地质时期 90
age of the Earth 地球的年龄 25—26
albedo of the Earth 地球的反照率 56
Aldebaran (star) 毕宿五(恒星) 18, 20
ALH84001 meteorite ALH84001 陨石 98—99
Alpha-Centauri B 半人马座 α 星 B 112
alteration minerals 蚀变矿物 93—94
aluminium-26 atoms 铝-26 原子 22—33
Ames Research Center 埃姆斯研究中心 1
amino acids 氨基酸 76
ammonia (NH_3) 氨 12, 107—109
Andromeda galaxy 仙女座星系 15
anti-biosignatures 反生命信号 119
Apex Chert rock formation, Australia 埃佩克斯燧石岩层，澳大利亚 42
archaea 古菌 66—69, 76, 78, 80
Archaean aeon 太古宙 32, 41, 46—48, 50—51
Aristotle 亚里士多德 3
asteroid impacts 小行星撞击 2, 24, 25, 30, 61—62, 96, 126; 参见 meteorites
astrometry 天体测量 111

atmosphere 大气 84—85
　on Earth 地球上的 28, 32—34, 44—53, 54, 55, 118—119, 125
　on exoplanets 系外行星上的 59, 117—119
　on Mars 火星上的 84, 85, 89—90, 94—96, 97
　on Pluto 冥王星上的 108
　on Titan (moon of Saturn) 土卫六上的 103—104
　on Triton (moon of Neptune) 海卫一上的 107
　on Venus 金星上的 84, 85—88
atoms common to life 构成生命的常见原子 9—10
ATP (adenosine triphosphate) 三磷酸腺苷 35

B

bacteria 细菌 66—69, 75—76, 78, 98; 参见 cyanobacteria
　survival in extreme heat 在极热条件下生存 79—80
banded iron formations 条带状铁建造 50
Bernal, J. Desmond J. 德斯蒙德·贝尔纳 6
Betelgeuse 参宿四 19
Big Bang 大爆炸 14, 16—17
bioastronomy 生命天文学 5—6
biomarkers 生物标志物 42—43, 54
biomass 生物量 49, 63—64, 102, 108
biomolecules 生物分子 70—71, 73

chirality 手性 37—39
biosignatures 生命信号 118—119
biospheres 生物圈 60, 63—65, 115, 121
black dwarfs 黑矮星 19
black holes 黑洞 20
body structures 身体结构 60
'boring billion' "枯燥的十亿年" 53—55
brain mass, proportional 脑化指数 122—123
Brasier, Martin 马丁·布拉西耶 42
Brock, Thomas 托马斯·布罗克 79—80
Brownlee, Don 唐·布朗利 125
Budyko, Mikhail 米哈伊尔·布德科 56
Buick, Roger 罗杰·别克 40

C

Callisto (moon of Jupiter) 木卫四 102
Cambrian Explosion 寒武纪生命大爆发 54, 59, 60
cap carbonates 盖帽碳酸岩 57
carbohydrates 碳水化合物 71, 78
carbon, organic 有机碳 10, 33—36, 39—40, 42, 50—52, 55, 74
carbon assimilation experiment 碳同化实验 96
carbon-based life 碳基生命 9—10
carbon dioxide 二氧化碳 11, 32, 34
　　in the Earth's atmosphere 地球大气中的 45—47, 51, 57
　　greenhouse effect 温室效应 45—48, 85—86, 95, 116
　　in the habitable zone 宜居带中的 117, 129
　　on Mars 火星上的 84, 85, 89, 94—96
　　on Venus 金星上的 84, 85, 86
carbonate-silicate cycle 碳酸盐-硅酸盐循环 47—48
Cassini-Huygens mission "卡西尼-惠更斯"计划 104—105
cells 单体 58, 65—70, 78
circulation 对流 8
Ceres (largest asteroid) 谷神星(最大的小行星) 83, 99
Charon (moon of Pluto) 冥卫一 108
chemical elements 化学元素 65
chemical weathering 化学风化 93
chemiosmosis 化学渗透 35
chemoheterotrophs 化能异养生物 74—75
Chicxulub, Mexico 希克苏鲁伯,墨西哥 61—62
chirality 手性 37—39
Chiron 喀戎星 29—30
chlorine 氯气 58—59
chromosomes 染色体 69—70
Chyba, Christopher 克里斯托弗·希巴 9
Cleland, Carol 卡罗尔·克莱兰 9
climate 气候
　　on Mars 火星上的 94—96
　　regulation 调节 47—48
CO_2 见 carbon dioxide
Cocconi, Guiseppe 朱塞佩·科科尼 120
coherent energy 相干能量 6—7
continents, formation 大陆形成 28—29

continuously habitable zone (CHZ) 连续宜居带 115

convergent evolution 趋同演化 122

Copernicus Principle 哥白尼原则 125—126

coronograph 日冕仪 113

Cosmic Connection (Sagan)《宇宙的连接》(萨根) 13

Cosmic Microwave 宇宙微波 Background 背景辐射 16—17

cosmobiology 宇宙生物学 6

Cosmotheoros (Huygens)《被发现的天上的世界》(惠更斯) 4

Creataceous-Paleogene mass extinction 白垩纪–第三纪大灭绝 61—62

Crick, Francis 弗朗西斯·克里克 72

cryovolcanism 冰火山 107—108

Curiosity Rover "好奇号" 火星车 89—90

cyanobacteria 蓝细菌 42—43, 49—50, 68—69, 76

D

Darwin, Charles 查尔斯·达尔文 8—9, 31, 126

Dawn mission "黎明号" 计划 99

definition of astrobiology 天体生物学的定义 1—2, 5

Democritus 德谟克里特 3

density of exoplanets 系外行星密度 114—115

direct detection of exoplanets 系外行星的直接探测 113—115

dissipative structures 耗散结构 8

DNA 脱氧核糖核酸 32—33, 36, 69—70, 71—73

DNA sequencing DNA 序列 77—78

Doppler shift 多普勒频移 111—112

Drake, Frank (Drake Equation) 弗兰克·德雷克(德雷克方程) 120—122

dropstones 坠石 56

dust particles 尘埃颗粒 10

dwarfs 矮星 19—21, 106, 116—117

E

Earth 地球 84

　age 年龄 25—26

　atmosphere 大气 28, 32—34, 44—53, 54, 55, 118—119, 125

　basis of life 生命的基础 9, 12, 63—70

　earliest aeon 最早的时期 28—36

　development of intelligent life 智慧生命的出现 122, 124

　mass extinctions 大灭绝 60—62

　origin of life 生命的起源 1—2, 31—43

　as a 'Pale Blue Dot' "暗淡蓝点" 般的 118

　position 位置 14—15, 26—27

　Snowball Earth hypothesis "雪球地球"假说 55—58

　uniqueness 唯一性 3—4, 125—127

Ediacarans 埃迪卡拉生物群 54

Einstein, Albert 阿尔伯特·爱因斯坦 113

elements 元素

　chemical 化学的 65

　common to life 构成生命的常见的 9—10, 89

isotopes 同位素 22—23, 25—26
non-metallic 非金属 12
in stars 恒星中的 17—18, 19, 21, 117
enantiomers 对映异构体 38—39
Enceladus (moon of Saturn) 土卫二 103, 128
Encephalization Quotient (EQ) 脑化指数 122—123
endosymbiosis 内共生 69
energy 能量 6—9, 87, 106
 infrared 红外线的 45
 life-giving 维持生命的 65
 metabolic 新陈代谢的 35, 58, 73—75
 in photons 光子中的 18
 precursor for life 生命存在的前提 58
entropy 熵 6—8
eukaryotes 真核生物 59, 66—70, 76, 78, 80
Europa (moon of Jupiter) 木卫二 100—102, 128—129
 ice on 其上的冰 2—3
European Space Agency 欧洲航天局 114
evolution 演化 8—9, 48, 75
 during 'boring billion' 在"枯燥的十亿年"中 53—55
 chemical 化学的 31
 convergent 趋同的 122
 diversity in 其中的多样性 60
exobiology 太空生物学 5
exoplanets 系外行星 3, 24—25
 atmosphere on 其上的大气 59, 117—119
 detection 探测 110—115
 evidence of life on 其上存在生命的证据 115—119
 intelligent life on 其上的智慧生命 120—124
expansion of the universe 宇宙膨胀 16—17
extinctions, mass 大灭绝 60—62
extraterrestrial intelligence (SETI) "搜寻地外文明"计划 120—124
extraterrestrial life 地外生命
 evidence of 其证据 2—3, 4
 intelligent 智慧 120—124
 likelihood of carbon base 碳基的可能性 9—10
 likelihood of silicon base 硅基的可能性 10—11
 probability 可能性 16
 significance 重要性 12—13
 theories about 其理论 3—4
extremophiles 极端环境微生物 79—81

F

faint young Sun paradox 暗淡太阳悖论 44—45
Fermi, Enrico (Fermi Paradox) 恩里科·费米（费米悖论）123—124
fluorine 氟 58—59
fossils 化石 41—42, 53—54, 98

G

G-type stars G 型恒星 125—126
galactic filaments 宇宙纤维状结构 16
galactic habitable zone (GHZ) 星系

宜居带 15, 117—118, 126
galaxies 星系 15—16
 birth of 其诞生 17
Galilean moons of Jupiter 木星的伽利略卫星 99—102
Galileo 伽利略 126
Galileo spacecraft "伽利略号"探测器 2—3, 101, 118
gametes 配子 69—70
Ganymede (moon of Jupiter) 木卫三 102
gas chromatograph mass spectrometer (*Viking* lander) 气相色谱质谱仪("海盗号"着陆器) 97
gas exchange experiment (*Viking* lander) 气体交换实验("海盗号"着陆器) 96—97
gas giants 气态巨行星 23—24
gene transfers 基因转移 77, 79
genetics 基因 75—79
genomes 基因组 8, 36—37
giant impact hypothesis "大碰撞"假说 26
giant planets 巨行星 23—24
glaciations 冰川作用 55—58
Great Oxidation Event 大氧化事件 48—53
greenhouse effect 温室效应 45—48, 85—86, 116
 on Mars 火星上的 95
gullies on Mars 火星上的冲沟 91

H

habitable zone (HZ) 宜居带 115—117
Hadean aeon 冥古宙 28—36

Haldane, J. B. S. J. B. S. 霍尔丹 32
half-life 半衰期 22, 25
Hawking, Stephen 斯蒂芬·霍金 126
helium 氦 17—18, 20
Herrmann, Joachim 约阿希姆·赫尔曼 5
Hertzsprung-Russell (H-R) diagram 赫罗图 20—21
Holmes, Arthur 亚瑟·霍姆斯 25—26
homochirality 同手性 39
Hooker, Joseph 约瑟夫·胡克 31
hot Jupiters 热木星 24—25, 110, 112
Hubble, Edwin 埃德温·哈勃 16
Huntress, Wesley 韦斯利·亨特里斯 5
Huygens, Christiaan 克里斯蒂安·惠更斯 4, 104—105
hydrocarbons on Titan 土卫六上的碳氢化合物 106
hydrogen 氢 17, 20, 52—53, 88
hydrogen bombs 氢弹 18
hydrogen peroxide 过氧化氢 90
hydrothermal vents 热液喷口 34—35, 42

I

ice 冰
 density 密度 11—12
 on Europa (moon of Jupiter) 木卫二上的 2—3
 formation 形成 55—57
ice giants 冰态巨行星 23—24
impact erosion 撞击侵蚀 96
impacts with Earth 撞击地球 28—30,

33—34, 61—62

indirect detection of exoplanets 间接系外行星探测 110—113

infrared radiation 红外辐射 45—46, 87, 93

inner planets, water on 内行星上的水 84

intelligent life on exoplanets 系外行星上的智慧生命 120—124

interferometry 干涉测量法 114

intraterrestrial life 地内生命 64

Io (moon of Jupiter) 木卫一 100

iron 铁 50, 51

iron-60 atoms 铁-60 原子 23

isotopes 同位素 22—23

    radioactive 放射性的 25—26

Isua, Greenland 伊苏亚，格陵兰岛 39—40

J

Jupiter 木星 65, 85

    formation 形成 24

    moons 卫星 2—3, 99—102, 128—129

    orbit 轨道 111

    as protection for Earth 保护地球 126—127

    resonance with Saturn 与土星共振 30

K

K stars K 型恒星 116

Kant, Immanuel 伊曼努尔・康德 4, 22

Kelvin scale 开尔文温标 20

Kepler, Johannes 约翰尼斯・开普勒 3—4

Kepler mission "开普勒"计划 112—113, 126

Kirschvink, Joe 乔・克里什芬克 57

Klein, Harold 'Chuck' 哈罗德・"查克"・克莱因 97

Kuiper Belt objects (KBOs) 柯伊伯带天体 107, 108

L

labelled release experiment 标记释出实验 97

Lafleur, Laurence 劳伦斯・拉弗勒 4—5

Lake Vostok, Antarctica 沃斯托克湖，南极洲 80—81

Laplace, Simon-Pierre (Marquis de Laplace) 西蒙-皮埃尔・拉普拉斯（拉普拉斯侯爵）22

Laplace resonance 拉普拉斯共振 100

Late Heavy Bombardment 晚期重轰炸 30, 39

lead isotopes, age 铅同位素的年龄 25—26

Lederberg, Joshua 乔舒亚・莱德伯格 5

life 生命

    characteristics 特征 6—9

    origin on Earth 在地球上的起源 31—43

precursors 前提 58—59, 65
light 光
    bending 弯曲 113
    cancelling 消除 114
light years 光年 15
lightning, creation of organic molecules 闪电制造有机分子 32
Linnaeus, Carolus 卡洛斯·林奈 66
lipids 脂类 71
liquid water 液态水 10, 11—12, 65, 82—86, 89, 90—96, 99—103, 106, 115—116, 119
Local Group 本星系群 15
Lowell, Percival 帕西瓦尔·罗威尔 4

M

M dwarfs M 型矮星 116—117
MacGregor, Alexander 亚历山大·麦格雷戈 32
magnesium-26 atoms 镁-26 原子 22—23
magnetic field 磁场 56
    on Europa 木卫二上的 101
magnetite 磁铁矿 56, 98—99
main sequence 主序 20—21, 44, 115—116
Mariner missions to Mars 探索火星的"水手"计划 88—89
Mars 火星 84—86, 88—99
    in the habitable zone 在宜居带中 116
    intelligent life 智慧生命 4
    signs of ancient life 远古生命迹象 2
Mars Exploration Rovers "火星探测漫游者"计划 94
Mars Science Laboratory 见 Curiosity Rover
mass extinctions 大灭绝 60—62
Mayor, Michel 米歇尔·梅厄 111—112
medium for biochemical processes 生物化学过程的介质 11
Mercury 水星 84—85
messenger RNA (mRNA) 信使 RNA 73
metabolism 新陈代谢 35, 58, 73—75
metallicity of stars 恒星的金属丰度 117—118
meteorites 陨石 25, 33—34; 参见 asteroid impacts
    age 年龄 26
    from Mars 火星上的 97—99
methane 甲烷 55, 90
    atmospheric 大气中的 46—47
    on Titan 土卫六上的 104—106
Methanopyrus kandleri 甲烷嗜热菌 80
Metrodorus 迈特罗多鲁斯 3
microbes 微生物 64—65, 127
    extraterrestrial 地外的 2
    gene transfer 基因转移 77, 79
    survival in extreme heat 在极端高温下生存 29, 79—80
microbial conjugation 微生物的接合 69
microbial mats 微生物席 40—41
microfossils 微体化石 41—42, 98
microlensing 微引力透镜 113
Milky Way galaxy 银河系 15, 16

Miller, Stanley 斯坦利·米勒 32—33
Mitchell, Peter 彼得·米歇尔 35
molecular clocks 分子钟 78—79
molecules, organic 有机分子 42—43
Moon 月球 14
    craters 陨石坑 30
    formation 形成 26—27, 28
Morrison, Philip 菲利普·莫里森 120
Murchison meteorite 默奇森陨石 33

N

NASA 美国国家航空航天局 1, 2, 5, 88—89, 99, 101, 112, 114, 118, 126
natural selection 自然选择 8, 75
Nazca plate, South Pacific 纳斯卡板块, 南太平洋 47
nebular hypothesis "星云"假说 22—23, 25
negative entropy 负熵 7
Neoproterozoic glaciations 新元古代冰川作用 55, 56, 57
Neptune 海王星
    distance 距离 15
    formation 形成 24
    moons 卫星 106—108
    orbits 轨道 30
neutrons 中子 19
Nice model 尼斯模型 30
non-metallic elements 非金属元素 12
North Pole, Australia 诺思波尔, 澳大利亚 40—41
nuclear fusion 核聚变 17—18, 19, 21
nucleotides 核苷酸 71—73

O

oceanic plates 大洋板块 47
Of the Plurality of Worlds (Whewell)《世界的多样性》(休厄尔) 4
Oort Cloud 奥尔特星云 126
Oparin, Alexander 亚历山大·奥帕林 32
organic carbon 有机碳 10, 33—36, 39—40, 42, 50—52, 55, 74
organic molecules 有机分子 32—36
origin of life on Earth 生命在地球上的起源 31—43
Orion Arm of the Milky Way 银河系猎户座旋臂 15
outflow channels on Mars 火星上的溢流河道 91—93
oxygen 氧 32
    on Europa 木卫二上的 102
    on exoplanets 系外行星上的 127
    levels 水平 48—53, 54, 55
    precursor for life 生命存在的前提 58—59
oxygenation time 氧化时间 59
ozone layer 臭氧层 52

P

PAHs (polycyclic aromatic hydrocarbons) 多环芳烃 98
'Pale Blue Dot'《暗淡蓝点》118
Paleoproterozoic glaciations 古元古代冰川作用 55, 56, 57
panspermia 生物外来论 31

Patterson, Clair 克莱尔·帕特森 26
Pauling, Linus 莱纳斯·鲍林 7, 78
PCR (polymerase chain reaction) technology 聚合酶链式反应技术 80
Pelagibacter ubique 遍在远洋杆菌 64—65
Permian-Triassic mass extinction 二叠-三叠纪大灭绝 61
Phanerozoic Aeon 显生宙 54
photons in the Sun 太阳中的光子 18
photosynthesis 光合作用 49—51
phylogeny 系统发育 77—79
Pikaia 皮卡虫 54
planetary embryos 行星胚胎 24
planetary migration 行星迁移 25
planetary nebulae 行星状星云 18—19
planetesimals 星子 24
plate tectonics 板块构造学 48, 127
Plato 柏拉图 3
pluralism 多元主义 3
Pluto 冥王星 108—109
primeval lead 原始铅 26
primordial soup 原始汤 32
prokaryotes 原核生物 67—68
protein synthesis 蛋白质合成 77
proteins 蛋白质 71, 76
Proterozoic Aeon 元古宙 55
proton gradient 质子梯度 35
Proxima Centauri 比邻星 15

Q

Queloz, Didier 迪迪埃·奎洛兹 111—112

R

racemic mixture 外消旋混合物 38
radial velocity method 径向速度法 111
radioactive isotopes 放射性同位素 25—26
radiogenic heat 放射性产热 101
Rare Earth Hypothesis "稀有地球"假说 125—127
recombination 重组 70
red dwarfs 红矮星 21
red giants 红巨星 18
red supergiants 红超巨星 19
redox titration 氧化还原滴定 53
reductants 还原剂 52
ribosomes 核糖体 68
RNA in (rRNA) 核糖体中的 RNA（核糖体 RNA）77
RNA 核糖核酸 36—37, 39, 71—73, 77
rocky planets 岩质行星 21, 23
runaway greenhouse effect on Venus 金星上失控的温室效应 86—87
runaway limit 失控极限 87—88

S

Sagan, Carl 卡尔·萨根 13, 14, 127
sample return missions 采样返回任务 128—129
Saturn 土星
  formation 形成 24
  moons 卫星 82, 102—106, 129
  orbit 轨道 111
  resonance with Jupiter 与木星共振 30

scablands 疤地 93
scale of the universe 宇宙的尺度 14—16
Schiaparelli, Giovanni 乔范尼·夏帕雷利 4
Schneider, Eric 埃里克·施耐德 8
Schopf, Bill 比尔·朔普夫 42
Schröinger, Erwin 欧文·薛定谔 7, 8
Second Law of Thermodynamics 热力学第二定律 7
Secret Service 特勤局 1
sedimentary rocks 沉积岩 39—41, 50
SETI (search for extraterrestrial intelligence) "搜寻地外文明" 计划 120—124
sexual reproduction 有性繁殖
    early 早期 54
    of eukaryotes 真核生物的 69—70
Shapley, Harlow 哈洛·沙普利 115
silicon based life 碳基生命 10—11
Simpson, George Gaylord 乔治·盖洛德·辛普森 5
size of Mars 火星的体积 96
Snowball Earth hypothesis "雪球地球"假说 55—58
social animals 社会性动物 123
soil on Mars 火星土壤 96—97
solar flux 太阳辐射通量 85, 87
Solar System 太阳系 15
    age 年龄 25—26
    formation 形成 22—25
    planetary orbits 行星轨道 30—31
    possible abodes of life 可能存在生命的天体 83—109, 128—129
South Africa, microfossils 南非, 微体化石 42
species 物种 75—76
spectra of exoplanets 系外行星光谱 119
spectral class 光谱型 21
SPONCH elements 硫、磷、氧、氮、碳、氢元素 12
stable medium for biochemical processes 生物化学过程的稳定介质 11
stars, lifecycle 恒星的生命周期 18—21
Strelley Pool Formation, Australia 斯特雷利池岩层, 澳大利亚 42
stromatolites 叠层石 40—41, 50
Struve, Otto 奥托·斯特鲁维 5
subduction 板块俯冲 47
subsurface biosphere 深部生物圈 64
sulphates on Mars 火星上的硫酸盐 94
sulphur 硫 51—52
sulphur dioxide on Mars 火星上的二氧化硫 95
Sun 太阳 14—15, 44—45
    elements in 其中的元素 17—18
    lifecycle 生命周期 18—19, 21
Super-Earths 超级地球 110
supernovae 超新星 19—20, 21, 23
surface stability 表面稳定性 65
symmetry 见 chirality
synchronous rotation 同步自转 116—117

T

taxonomic levels 分类级别 66
technosignatures 技术信号 120—124
telescopes, space-based 天基望远镜

113—114

Thales 泰勒斯 3

Theia, impact with Earth 忒亚撞击地球 26

Theory of General Relativity 广义相对论 113

thermodynamics 热力学 6—8

thermophiles 嗜热菌 29—30, 79—81

Thermus aquaticus 水生栖热菌 80

tidal locking 潮汐锁定 116—117

Tikov, Gavriil 加夫里尔·季科夫 5

tillites 冰碛岩 56

tilt 轴倾角 127

    of Mars 火星的 95

Titan (moon of Saturn) 土卫六 82, 103—106, 129

transcription 转录 73

transit method 凌日法 112—113, 119

transit timing variations 凌日时间变分 114

Transiting Exoplanet Survey Satellite (TESS) 凌日系外行星巡天卫星 113

transmitting civilizations 射电文明 120—124

tree of life 生命之树 76, 77

Triton (moon of Neptune) 海卫一 106—108

Tumbiana Formation, Australia 坦比亚纳组, 澳大利亚 50

U

universe 宇宙

    expansion 膨胀 16—17

    size 大小 14—16

uranium isotopes, age 铀同位素的年龄 25

Uranus 天王星

    formation 形成 24

    orbits 轨道 30

Urey, Harold 哈罗德·尤里 32—33

V

valley networks on Mars 火星上的河谷网络 91—92

Venus 金星 84—88

Viking landers "海盗号"着陆器 9, 89, 96—97

Virgo supercluster 室女座超星系团 15

viruses 病毒 70

volcanoes 火山 33, 57

    on Io 木卫一上的 100

    cause of mass extinction 大灭绝的原因 61

Voyager 1 spacecraft "旅行者1号"探测器 118

W

Ward, Peter 彼得·沃德 125

water, liquid 液态水 10, 11—12, 65, 82—86, 89, 90—96, 99—103, 106, 115—116, 119

Waterbelt Earth "水带地球" 57—58

Watson, James 詹姆斯·沃森 72

weird life 怪异生命 82, 106

Whewell, William 威廉·休厄尔 4,

115

white dwarfs 白矮星 19

wobbles in stellar orbits 恒星轨道摆动 111

Woese, Carl 卡尔·乌斯 76—77

Z

zircons 锆石 29

Zuckerkandl, Emile 埃米尔·祖克坎德尔 78

David C. Catling

ASTROBIOLOGY

A Very Short Introduction

Contents

Acknowledgements i

List of illustrations iii

1 What is astrobiology? 1

2 From stardust to planets, the abodes for life 14

3 Origins of life and environment 28

4 From slime to the sublime 44

5 Life: a genome's way of making more and fitter genomes 63

6 Life in the Solar System 82

7 Far-off worlds, distant suns 110

8 Controversies and prospects 125

Further reading 131

Acknowledgements

I thank Professor John Armstrong, Dr Rory Barnes, Professor John Baross, Dr Billy Brazelton, Professor Roger Buick, Dr Rachel Horak, Imelda Kirby, and Professor Woody Sullivan for reading parts of the manuscript and offering various corrections and suggestions. Thanks also to Latha Menon and Mimi Southwood, who read the entire manuscript, the latter offering the perspective of a non-scientist. Numerous students who attended my astrobiology classes at the University of Washington over the years also helped me mentally prepare for writing this book. At Oxford University Press, I thank Emma Ma and Latha Menon for their encouragement and assistance.

List of illustrations

1 The Hertzsprung–Russell diagram **20**
Adapted from 'Stellar Evolution and Social Evolution: A Study in Parallel Processes' (2005), *Social Evolution & History*. 4: 1, 136–59), reproduced with permission of Professor Robert Carneiro

2 *Left*: Cross-section of the world's oldest fossil stromatolites *Right*: A plan view of the bedding plane of the stromatolites **40**
Photographs by David C. Catling

3 The approximate history of atmospheric oxygen **49**
Author's own diagram

4 a) Schematic of prokaryote (archaea and bacteria) versus eukaryote structure; b) Two bacteria caught in the act of conjugation **67**
b) Credit: Charles C. Brinton Jr. and Judith Carnahan

5 *Left*: DNA consists of two strands connected together. *Right*: In three dimensions, each strand is a helix, so that overall we have a 'double helix' **72**

6 The classification scheme for metabolisms in terrestrial life **74**

7 The 'tree of life' constructed from ribosomal RNA **76**

8 a) Valley networks on Mars; b) Outflow channel Ravi Vallis **92**
a) ESA/DLR/FU Berlin (G. Neukum); b) NASA/JPL/Caltech/ Arizona State University

9 The Galilean moons of Jupiter: Io, Europa, Ganymede, and Callisto **100**
NASA/JPL/DLR

10 a) A network of channels that appear to flow into a plain near the Huygens landing site; b) Image of the surface at the Huygens landing site. **105**
Courtesy of ESA/NASA/University of Arizona

11 Brain and body mass for some different mammals. **123**
Adapted from Comparative Biochemistry and Physiology Part A: Molecular & Integrative Physiology, Vol. 136: 4, Hassiotis, M., Paxinos, G., and Ashwell, K. W. S., 'The anatomy of the cerebral cortex of the echidna (Tachyglossus aculeatus)', 827–50. Copyright (2003), with permission from Elsevier

Chapter 1
What is astrobiology?

Behind the name

'What the hell is *astrobiology*?' an American Secret Service agent cried into his walkie-talkie. He had just been checking the identity of an academic visitor to NASA's Ames Research Center, near San Francisco. The visitor had said that he was attending NASA's first astrobiology science conference. Ames has an airstrip that provides a secure landing site for Air Force One, and, in April 2000, President Bill Clinton had just flown in to visit the San Francisco Bay area, bringing along his Secret Service entourage.

The agent's question was a fair one. It was only in the late 1990s that a scientific consensus emerged about the meaning of *astrobiology*. Few laymen or Secret Service agents would have heard of the term. Back then, NASA began to promote a research programme in astrobiology led by Ames, where I was working as a space scientist. At first, some of my colleagues disliked the literal Greek meaning of the 'biology of stars'. One noted with a scoff how life couldn't exist inside the infernos of stars. A less curmudgeonly interpretation is that the 'astro' in astrobiology concerns life *around* stars, including the Sun, or simply life in space. In fact, many astrobiologists are as much concerned with the history of life on Earth as with life elsewhere. Astrobiologists agree that we should have a firm understanding of how life evolved on Earth in order to ponder the existence of life in outer space. Yet one of the astonishing aspects of modern science is that it has so far failed to answer questions about biology that even a child might ask. How did life on Earth get started? We have some ideas but the details are unknown. Which special properties of the Earth and the Solar System make our planet habitable? Again, some thoughts but there is still much to learn. And what caused life to evolve into complex organisms instead of remaining simple? Again, we're uncertain.

To fill these holes in human knowledge, astrobiology has emerged as *a branch of science concerned with the study of the origin and evolution of life on Earth and the possible variety of life elsewhere*. This is my own preferred definition. NASA has defined astrobiology as *the study of the origins, evolution, distribution, and future of life in the universe*. Other common definitions are *the study of life in the universe* or *the study of life in a cosmic context*. Within this purview, astrobiologists pursue the question 'What's the history and future of terrestrial life?' as well as 'Is there life elsewhere?'

Four developments coincided with the emergence of astrobiology as a discipline in the late 1990s. In 1996, controversial signs of ancient life were described within a Martian meteorite—a 1.9 kilogram piece of rock that had been blasted off the surface of Mars by an asteroid impact and had eventually landed in Antarctica. Whether the interpretation of fossilized microscopic life was correct or not (see Chapter 6), it set people thinking. Furthermore, over the preceding two decades, biologists had established that some microbes not only tolerated a much larger range of environments than had previously been thought but actually *thrived* in extremes of temperature, acid, pressure, or salinity. So it became plausible to contemplate extraterrestrial microbes existing in seemingly hostile places. A third finding came in 1996 from pictures taken by NASA's *Galileo* spacecraft of the ice-covered surface of Jupiter's moon, Europa, which revealed pieces of ice that had drifted apart in the past, suggesting an ocean below an icy crust. Then, from the mid 1990s onwards, astronomers found increasing numbers of extrasolar planets or *exoplanets*, which are planets orbiting not our Sun but other stars. The possibility that life might reside on exoplanets or in the cosmic backyard of our own Solar System provided an impetus for asking whether life might be common in the universe.

Astrobiology in the history of ideas

Although astrobiology came to the fore in the 1990s, the question of whether we're alone in the universe goes back millennia. Thales (*c.* 600 BC), often regarded as the father of Western philosophy, espoused the idea of a *plurality of worlds* with life. Subsequently, the Greek atomist school from Leucippus to Democritus and Epicurus, which believed that matter was made of indivisible atoms, favoured such 'pluralism'. Metrodorus (*c.* 400 BC), a follower of Democritus, wrote, 'It is unnatural in a large field to have only one stalk of wheat, and in the infinite universe only one living world'. But it would be incorrect to equate the ancient philosophers' pluralism with our modern conception of life on Mars or exoplanets. Metrodorus had no clue that stars were Sun-like objects at enormous distances and believed that they formed daily from moisture in the Earth's atmosphere. The populated worlds of the atomists' imagination were bodies in an intangible space, similar to modern ideas of parallel universes. In any case, the opposing view of Plato (427–347 BC) and Aristotle (384–322 BC) ultimately dominated. Their belief that the Earth was uniquely inhabited and at the centre of the universe prevailed for over a thousand years.

Eventually, Renaissance astronomers showed that the Earth orbited the Sun. With the realization that the Earth was merely another planet, speculations soon arose about extraterrestrial life on other planets in the Solar System. Johannes Kepler (1571–1630), the German astronomer responsible for astronomy's three laws of planetary motion, happily entertained the idea of inhabited planets. Then, by the end of the 17th century, the Dutch astronomer Christiaan Huygens (1629–95) was imagining life *beyond* the Solar System in his book *Cosmotheoros* (1698): 'all those Planets that surround that prodigious number of Suns. They must have their plants and animals, nay and their rational creatures too.' Extraterrestrial life was so in vogue that in 1755 the philosopher Immanuel Kant (1724–1804) wrote of intellectuals on Jupiter and amorous Venusians.

In contrast, some scholars with religious views continued to cling to Earth's uniqueness. An example was Cambridge University's William Whewell (1794–1866), whose *Of the Plurality of Worlds* (1853) argued against other inhabited planets in a sort of forerunner of a contemporary debate called the *Rare Earth Hypothesis* that I discuss in Chapter 8.

By the late 19th century, the issue of life elsewhere was seen as a purely scientific matter, though science itself soon developed some blind alleys. Telescopic observations by Giovanni Schiaparelli (1835–1910) and Percival Lowell (1855–1916) created a surge of interest in the possibility of intelligent life on Mars. Unfortunately, Lowell's belief that he saw canals on Mars was an optical illusion created when the mind connects dots in blurry images, and his ideas of Martian civilizations were fantasy. Increasingly, as astronomers employed painstaking techniques such as examining spectra of light from planets, it became apparent that the physical conditions on various Solar System planets might not be so favourable for life after all. The pendulum swung so firmly in the other direction that by the mid 20th century few astronomers were interested in planets. It took the Space Age to rekindle old curiosities.

While the basic questions of astrobiology are ancient, the term 'astrobiology' only surfaced from time to time before becoming common in the 1990s. In 1941, an essay entitled 'Astrobiology' by

Laurence Lafleur (a philosopher at Brooklyn College, New York) described the word more narrowly than its modern incarnation as the consideration of life other than on Earth. Otto Struve, an astronomer at the University of California in Berkeley, also used the term in 1955 to describe the search for extraterrestrial life. A Russian astrophysicist, Gavriil Tikov (1875–1960), and a German astronomer, Joachim Herrmann, published books entitled *Astrobiology* in 1953 and 1974, respectively, covering popular ideas of extraterrestrial life.

The modern use of 'astrobiology' was introduced in 1995 by Wesley Huntress, then at NASA's headquarters in Washington DC. At the time, NASA scientists argued that a study of life over scales from the microbial to the cosmic was essential for understanding life in the universe. Huntress liked the word *astrobiology* for this aspiration and the name stuck.

In fact, astrobiology was really a reinvention and expansion of *exobiology*, a field that goes back several decades. In 1960, Joshua Lederberg (1925–2008) coined the word exobiology for 'the evolution of life beyond our own planet'. Lederberg, a Nobel Prize winner for discoveries in bacterial genetics, argued that an essential part of space exploration should be searching for life. Then, from the 1960s onwards, NASA followed Lederberg's advice and financed exobiology research. But exobiology soon developed its critics. In 1964, George Gaylord Simpson, a Harvard biologist, quipped that 'this "science" has yet to demonstrate that its subject matter exists'.

The advantage of today's astrobiology is that it cannot be held to Simpson's charge because it includes the study of the origin and evolution of life on Earth at its core. Astrobiology also emphasizes the origin and evolution of planets as a context for life, and so embraces astronomers more firmly than exobiology.

Exobiology is not the only term similar to astrobiology. Since 1982, astronomers have officially used *bioastronomy* for the astronomical aspects of the search for extraterrestrial life, while earlier the word *cosmobiology* was favoured by J. Desmond Bernal (1901–71), an influential Irish-born British physical chemist. But neither of those terms has become widespread.

What is life?

Astrobiology raises the difficult question of how to define life. What is it exactly that we are looking for beyond Earth? A common approach is to list life's characteristics, which include reproduction, growth, energy utilization through metabolism, response to the environment, evolutionary adaptation, and the ordered structure of cells and anatomy. Unfortunately, this way of defining life is unsatisfactory for a couple of reasons. First, the list describes what life does rather than what life is. Second, most of these aspects of life are not unique. Life has structural order such as cells, but salt crystals are also ordered. Some of my friends have no children but they're alive, I think, as are ligers that cannot reproduce. Growth and development apply to living entities, but also to spreading fires. All life metabolizes but so does my car. Life reacts to its environment, but a mercury thermometer also responds to its surroundings.

Alternatively, some scientists try to define life using *thermodynamics*, that is, heat and energy and their relationship to matter, by suggesting that the essence of life is the presence of stable structures, such as cells and genetic material, alongside *entropy* produced by metabolic waste and heat.

The term entropy requires some clarification. Desperate teachers searching for a quick and dirty explanation have often called it 'disorder'. Entropy is not disorder but an exact measure of energy dispersal amongst particles, be they atoms or molecules. Energy disperses spatially and so the energy of groups of particles that move together, which is said to be *coherent* energy, can dissipate. Thus, a bouncing ball comes to rest because its coherent energy of motion is converted into incoherent thermal motion of molecules and atoms through friction. In contrast, a stationary ball never spontaneously begins to bounce (as if it were alive) because even though sufficient thermal energy exists in the floor below, that energy is unavailable and dispersed in random jiggling of the atoms of the floor. The Second Law of Thermodynamics governs such phenomena and states that entropy in the universe never decreases. The entropy increase (or energy dispersal) conserves energy but ruins its quality. High-quality energy is not distributed but concentrated, such as in a barrel of oil, the nucleus of an atom, or in photons (particles of light) possessing high frequency and short wavelength. Such photons include ultraviolet and visible ones that cause sunburn and that power plant life, respectively. In physics, such high-quality energy has low entropy.

The physicist who most prominently linked entropy to life was the Nobel Laureate Erwin Schrödinger (1887–1961). In *What is Life?* (1944), Schrödinger commented that an organism 'tends to approach the dangerous state of maximum entropy, which is death. It can only keep aloof from it, i.e. alive, by continually drawing from its environment negative entropy… Indeed, in the case of higher animals we know the kind of orderliness they feed upon well enough, viz. the extremely well-ordered state of matter in more or less complicated organic compounds, which serve them as foodstuffs. After utilizing it they return it in a very much degraded form.' Regrettably, Schrödinger introduced the concept of 'negative entropy', which does not exist in science, to describe the ordered structure of food. Also, in the growth of some organisms, the increase in entropy primarily comes from heat generation rather than the degraded form of metabolic waste products compared to food. Linus Pauling (1901–94), who was arguably the greatest chemist of the 20th century, bluntly remarked that Schrödinger 'did not make any contribution whatever [to our understanding of life]… perhaps, by his discussion of "negative entropy" in relation to life, he made a negative contribution'.

Nonetheless a curious by-product of the ever-increasing entropy in the universe is that ordered, low-entropy structures, such as organisms, spring into existence. In fact, efficient production of entropy is best achieved by so-called *dissipative structures* involving a coherent *structure* of an immense number of molecules that *dissipates* energy. A simple example is a convection cell in boiling water. Warm water rises and is balanced by sinking water on its periphery. This circulating cell helps to disperse energy and so increases entropy more efficiently than if the cell were absent. All living organisms are complicated dissipative structures. However, attempts to define life with thermodynamics have so far failed to distinguish clearly between the living and non-living. For example, the writer Eric Schneider defines life as a 'far from equilibrium dissipative structure that maintains its local level of organization at the expense of producing environmental entropy'. A fire also fits this definition.

Pauling's criticism aside, Schrödinger argued correctly that organisms must run a sort of computer program, which is what we now call the *genome*. Indeed, life anywhere probably has to possess a genome. By a *genome*, we mean a heritable blueprint subject to small copying errors, which allows an organism to have evolved from an ancestor and provides a recipe for life's other characteristics such as metabolism. *Evolution*—the changes in populations over successive generations caused by selection of individuals' characteristics—is important because it is the only process that can explain the diversity of life and how the features of life that were listed previously were configured. In Darwin's natural selection mechanism, the genetic variation in populations of individuals means that some are better adapted for greater reproductive success than others. Natural selection favours genes that leave more descendants, so that lineages accumulate genetic adaptations.

Mindful of the centrality of evolution, astrobiologists often define life as 'a self-sustaining chemical system capable of Darwinian evolution'. Unfortunately, this definition is not helpful if we want to design an experiment to find life. Do we have to wait for evolution to happen for a positive detection? A better definition uses the past tense: 'life is a self-sustaining, genome-containing chemical system that *has* developed its characteristics through evolution'. So far, space-borne life detection experiments have not tried to measure a genetic make-up. For example, NASA's *Viking Lander* probes, which looked for life on Mars in the 1970s, were designed to recognize the Earth-like metabolism of microbes in the soil (Chapter 6).

The philosopher Carol Cleland and scientist Christopher Chyba have suggested that attempts to define life are like those of 17th-century scientists trying to define water. At that time, water was considered a colourless and odourless liquid that boils and freezes at certain temperatures. Without atomic theory, no one knew that water is a collection of molecules, each consisting of two atoms of hydrogen joined to an oxygen atom. By analogy, perhaps we lack the theory of living systems needed to define life.

Many of the problems in defining life boil down to the fact that we have only one example—life on Earth. All Earth-based organisms use nucleic acids for hereditary information, proteins to control biochemical reaction rates, and identical phosphorus-containing molecules to store energy. It's the same basic biochemistry in a bacterium or a blue whale. So it is difficult to distinguish which properties of life on Earth are unique and which are needed generally to qualify as 'life'. Astrobiology could help solve this conundrum if we found life beyond Earth.

The bare necessities of life

While there's no perfect definition of life, there are reasonable grounds to think that certain atoms common in terrestrial biochemistry are likely to be used by extraterrestrial organisms and might help us recognize life elsewhere. On Earth, the chief structural elements in biology are carbon, nitrogen, and hydrogen, while chemical interactions take place in liquid water. In astrobiology, there's wide agreement that life elsewhere is likely to be carbon based and that a planet with liquid water would, at least, favour 'life as we know it.' These deductions arise from realizing that life is constructed from a limited toolkit, the periodic table of chemical elements, which is the same throughout the universe.

In fact, carbon is the only element capable of forming long compounds of billions of atoms such as DNA (deoxyribonucleic acid). Consequently, only carbon-based extraterrestrial life seems able to have a genome of comparable complexity to terrestrial life. Carbon also has a variety of other properties that allow a unique chemistry of its compounds, sufficient to spawn the discipline of organic chemistry. Carbon's special properties include the ability to form single, double, and triple bonds with itself as well as bonds with many other elements. Carbon can also build three-dimensional complexity by forming hexagonal rings that join together.

Because life has to get started and propagate, it's probable that the main atoms of life are abundant ones. Carbon is fourth in cosmic abundance after hydrogen, helium, and oxygen. Indeed, astronomers have found that many non-biological organic molecules already exist in space. These freebies might serve as precursors to life getting started (see Chapter 3). For example, around 30 per cent by mass of the dust between the stars is organic material. So-called carbonaceous chondrite meteorites and interplanetary dust particles in our own Solar System contain up to 2 per cent and 35 per cent organic carbon by mass, respectively.

Because silicon has chemical properties similar to carbon, it is sometimes asserted that silicon might allow an alternative extraterrestrial biochemistry to carbon-based molecules, despite being about ten times less cosmically abundant than carbon. But in water, at least, silicon compounds tend to be unstable and silicon easily gets locked into solid silicon oxides. Carbon dioxide is a gas at common planetary temperatures and dissolves in water to concentrations sufficient for organisms to use carbon dioxide as a carbon source. Silicon dioxide, in contrast, is an insoluble solid, such as quartz. Silicon's bonds with oxygen and hydrogen are strong, whereas carbon–oxygen and carbon–hydrogen bonds are similar in strength to the carbon–carbon bond, which allows carbon-based compounds to undergo reactions of exchange and modification. Silicon–hydrogen bonds also tend to be easily attacked in water. The stability of silicon-based molecules requires low temperatures to slow down reactions that would otherwise destroy them. Appropriately cold solvents include oceans of liquid nitrogen on icy planets far from their stars. At present, such silicon-based life remains purely speculative.

However, a stable medium is necessary for biochemical processes such as metabolism or genetic replication; on Earth this medium is liquid water. For extraterrestrial life, the medium could be another liquid or a dense gas as long as it doesn't easily become a solid in the prevailing environment. Nonetheless, water (H_2O) has some unique properties. Unlike its smelly twin, hydrogen sulphide (H_2S), which condenses to a nasty liquid only at −61°C, water turns to liquid below 100°C at normal pressures. Liquid water's stability occurs because the oxygen atom in water molecules is slightly negatively charged and allows a relatively strong 'hydrogen bond' to the slightly positive hydrogen atoms of other water molecules. Sulphur provides weaker hydrogen bonds between hydrogen sulphide molecules. Water also forms stronger hydrogen bonds to other water molecules than to molecules of oily substances. As a consequence, oils separate from water, which allows cell membranes to form and provide homes for genes and metabolic processes.

Another unusual property of water is that ice is less dense than liquid water. When water freezes, the molecules align into ring-like structures containing open holes on the atomic scale. If ice were denser than liquid, the cold bottom of lakes and seas would collect ice, which would be insulated and remain frozen. Seas would freeze from the bottom up and become uninhabitable. This would arise because sunlight would be reflected back in the areas where the ice reached the surface, causing cooling and more ice to accumulate. Gradual freezing might be the unfortunate fate of seas of other liquids such as ammonia. Ammonia is a liquid from about −78°C to about −33°C at a pressure of one atmosphere. But any seas of ammonia would tend to solidify from the bottom up, unlike seas of water that remain liquid even if cold temperatures cause an ice cover.

On Earth, we find microbes wherever there's liquid water (excluding sterilized apparatus), so 'life as we know it' is as much 'water based' as 'carbon based.' Consequently, for Solar System exploration, the detection of liquid water or its past presence provides an objective for planetary probes, such as those visiting Mars. Nonetheless, it is possible to conceive of organic solvents as alternatives to water, which is potentially important for Titan, Saturn's largest moon (Chapter 6).

Another observation from terrestrial life is that just six non-metallic elements—carbon, hydrogen, nitrogen, oxygen, phosphorus, and sulphur—make up 99 per cent of living material by mass. These elements are often abbreviated as 'CHNOPS', but I prefer 'SPONCH', which is easier to say. We find H and O in water, which makes up most living tissue, and C, H, and O in the nucleic acids of genetic material and in carbohydrates. C, H, N, and S exist in proteins, while P is essential for nucleic acids and energy-storage molecules. Consequently, detecting the SPONCH elements in chemical forms that life could use is another practical goal for planetary space probes searching for Earth-like life.

The significance of life elsewhere

You might wonder whether the discovery of simple, microbial-like extraterrestrials would really matter. But if we could find a single instance of life that originated elsewhere, it would prove that life is not a miracle confined to Earth. We wouldn't be alone. Even the simplest microbes native to Mars or to Europa's oceans would change the balance of probabilities that life exists elsewhere in the galaxy for they would demonstrate that life can originate twice within one solar system. At the moment, we have no convincing evidence of life beyond Earth. However, in Chapter 6, I will argue that at least nine other bodies in our Solar System might be habitable today, if we keep an open mind. Solar System astrobiology is far from settled.

A second significant factor in finding life harks back to the difficulties of defining life. The planetary scientist Carl Sagan (1934–96) commented in his book *Cosmic Connection* (1973) that 'the science that has by far the most to gain from planetary exploration is biology'. An examination of extraterrestrial life would be of profound significance not only in identifying those elusive characteristics common to all life but also in shedding light on how life originated, which remains unsolved.

Chapter 2
From stardust to planets, the abodes for life

> To make an apple pie from scratch, you must first invent the universe.
>
> Carl Sagan (1980)

When the universe began, temperatures everywhere were far too hot for atoms to be stable, let alone join up into complex biological molecules. Life exists because following a *Big Bang* 13.8 billion years ago, a hot, dense cosmos expanded and cooled. As it did so, atoms, galaxies, stars, planets, and life arose. Here, we examine how this process produced an abode for life—the Earth.

We start with the structure of the present universe, which provides the clues about its history. Imagine a journey that goes to the edge of the observable universe. At the speed of light— 300,000 km per second—it would take only 1.3 seconds to get to the Moon at its distance of 384,000 km. A diagram of the Earth represented as a spot of 2.7 mm diameter and the Moon looks as follows:

Earth   Moon

This scale is fairly easy to grasp. But it challenges our imagination when we realize that the Sun on the same scale would be 30 cm in diameter and we would have to place it just over 30 m away from this book. The nearest star, Proxima Centauri, which is small compared to the Sun, would be about 4 cm in diameter. To keep to scale, we would have to place it 8,650 km away, which is roughly the flight path from San Francisco to London.

Continuing our voyage at the speed of light, it would take 8.3 minutes to fly from the Earth to the Sun and a little over four hours to then travel to the average orbital distance of Neptune, the outermost of the eight planets. After 4.2 years, we would reach Proxima Centauri. With the huge distances involved, we define the distance travelled in a year at the speed of light, some 9,500 billion km, as a *light year*, so that Proxima Centauri is 4.2 light years away.

The Sun and Proxima Centauri are two of about 300 billion stars in the Milky Way galaxy, which is a disc some 100,000 light years across with stars concentrated in spiral arms. Galaxies contain millions to trillions of stars, so the Milky Way is moderately large. The Solar System sits in the Orion Arm, two-thirds out from the Galactic Centre. This arm appears unremarkable compared to some others that are more richly populated with stars. But it's possible that the Solar System's location in the galactic sticks was actually vital for terrestrial life. The Earth may have avoided certain catastrophes, such as proximity to exploding stars. If so, there may be a particular region in galaxies favourable for life, called the 'Galactic Habitable Zone', discussed in Chapter 7.

On a larger scale, there are more than 100 billion galaxies in the observable universe arranged into groups and superclusters. Within a diameter of about 10 million light years centred between the Milky Way and our nearest spiral galaxy, the Andromeda galaxy some 2.5 million light years away, there are roughly fifty galaxies, forming the *Local Group*. In turn, this group is one of a hundred or so within a sphere of 110 million light years' diameter, comprising the *Virgo Supercluster*. Maps that cover billions of light years show filaments of tiny scattered points where each point is a galaxy. In three dimensions, galactic filaments, which are the largest structures known to humankind, join up into a web separated by vast voids. The whole fantastic structure looks as if it were spun by a crazy intergalactic spider.

How big is the observable universe? If space had not expanded, the farthest distance would be 13.8 billion light years, which is that traversed by a photon—a particle of light—since the Big Bang happened 13.8 billion years ago. But space has expanded. So the actual size of the observable universe is now about 47 billion light years across. Such vastness is a consideration for astrobiology because surveys of planets around stars just within the Milky Way suggest that each star hosts at least one planet on average. Some fraction of planets ought be habitable, perhaps at least 1 per cent, so the number of potential abodes for life might exceed a trillion billion.

The structure of the universe traces back to the Big Bang. In the 1920s, telescopic observations by the American astronomer Edwin Hubble showed that galaxies are moving away from each other on the large scale as space expands between them. Thus, going back far in time, everything must have been scrunched up and very hot. This consideration led to the recognition of strong evidence for the Big Bang. Its afterglow, the *Cosmic Microwave Background*, permeates the entire universe. If the Big Bang is true, physics dictates that before the early universe cooled, it must have been an opaque fireball made of electrons and protons (the electrically negative and positive elementary particles that make up atoms), photons (particles of light), and some small groups of fused protons. Some 380,000 years after the Big Bang, it became cool enough that electrons were able to join with protons, or groups of protons, to form the two smallest atoms, hydrogen and helium. At that point, the universe became transparent to light because previously photons had been scattered by the free-floating electrons. This prior situation was analogous to the way that light bounces around off tiny droplets in a fog so that you can't see through. Since losing its opaqueness, the universe has expanded by about a factor of 1,000, and the wavelength of the relic photons from the Big Bang has been stretched the same amount, changing the photons from red light to microwave. Amazingly, when you tune an old analogue television or radio between channels, the leftover radiation from the Big Bang contributes to the static hiss, albeit at a level of about 1 per cent or less. In fact, in 1964, this noise in a large radio horn was how the microwave background was discovered. Pigeon droppings in the horn were blamed initially; but after cleaning up and shooting the unfortunate pigeons, the real culprit was identified as the beginning of the universe.

Galaxies appeared a few hundred million years after the Big Bang. Some places had very slightly more material than average and so had higher gravity. Clumping produced galaxies, and within the galaxies, on a smaller scale, gas clouds collapsed under their own gravity. The interior of each shrinking cloud heated up as gas particles collided, eventually making a hot, glowing ball of gas—a star.

Part of the 'astro' in astrobiology comes from the fact that all of the atoms used by life except for hydrogen were created inside stars. The first stars would have been made only of the elements synthesized in the Big Bang: hydrogen, comprising three-quarters of the mass, and helium, which made up the rest except for a trace of lithium. The oxygen in water, the nitrogen in proteins, or the carbon in every organic molecule—none of these elements was present at first. But eventually stars made them.

To understand how stars make elements, consider how the Sun shines. Inside the Sun, immense heat strips each atom down into its constituents: a positively charged nucleus and negatively charged electrons. The temperature at the centre of the Sun, some 16 million degrees Celsius, is enough to fuse the nuclei of four hydrogen atoms into a helium nucleus, which is a nuclear reaction that releases photons. Each photon then endures a one-million-year journey from the interior of the Sun to space. It takes so long because each photon is continually absorbed and emitted as it encounters material. The photon also loses energy. It starts out as a high-energy gamma ray and on average turns into a lower-energy photon of visible light by the time it escapes from the Sun. In the 1950s, physicists reproduced the kind of nuclear reactions that occur inside the Sun with hydrogen bombs. Cores of stars like our Sun are akin to hydrogen bombs that, in effect, can't explode because they are contained by the weight of material above them.

The Sun will not fuse hydrogen in its core forever and the repercussions will destroy life on Earth (partly answering astrobiology's question of 'What's the future of life?'). In most stellar cores, the accumulation of helium 'ash' causes temperatures to drop too low to support further hydrogen fusion. At this point, the star shrinks under its own weight, which causes the temperature to rise until it ignites hydrogen fusion in a shell surrounding the core. The energy release causes outer layers of the star to expand, cool, and redden. This is how a *red giant* forms, such as Aldebaran, the brightest star in the constellation of Taurus, the Bull. The Sun will eventually become a red giant and swell two-hundredfold by 7.5 billion years' time, probably engulfing the Earth. The weight of further helium ash eventually squeezes a red giant's core to a temperature of 100–200 million degrees Celsius, which is enough to fuse helium nuclei and make carbon and oxygen. In turn, a helium-burning shell can eventually surround a core of carbon and oxygen 'ash'. Stars with four to eight times the mass of the Sun even end up fusing the carbon and oxygen into heavier elements, including neon and magnesium.

Generally, the death throes of a Sun-like star involve shedding outer layers into space. These shells of glowing gas are called *planetary nebulae* because they look somewhat like planets through low-magnification telescopes, but they have nothing to do with planets. The remnants of the star cool down into a *white dwarf* with a radius comparable to that of the Earth but an enormous density. In theory, a white dwarf stops shining after tens to hundreds of billions of years, producing a *black dwarf*. But the universe is not yet that old.

Stars larger than about eight times the mass of the Sun eventually explode as *supernovae*. The Sun is about half way through its ten-billion-year phase of hydrogen fusion in its core, but these massive stars spend less than 60 million years in the same phase before becoming *red supergiants*, of which Betelgeuse, in the constellation of Orion, is an example. In such stars, nuclear fusion in shells around the core produces the elements neon, magnesium, silicon, and iron. Iron is generally the heaviest element made, although some heavier elements are also produced when free neutrons are added to existing nuclei. (Neutrons are particles that have no electrical charge and are commonly found in atomic nuclei.) When the fuel in such a star runs out, the core is so compressed that negatively charged electrons amalgamate with the positively charged protons of the iron nuclei. The electrical charges cancel out and uncharged neutron particles are created. As a result, the stellar core shrivels to about 12 km diameter, forming something like a gigantic atomic nucleus made only of neutrons. Because of the shrinkage, the rest of the star collapses onto the dense neutron core, and a violent rebound creates the incredibly bright supernova.

Elements heavier than iron are generated and distributed by supernovae. A few seconds after a supernova, the outer layers of the star are heated to an incredible 10 billion degrees Celsius while the break-up of nuclei from deeper layers supplies abundant neutrons to fuel reactions that make heavy elements. This cosmic alchemy creates the heavy, precious elements such as gold, silver and platinum. Importantly, the material blown out in a supernova

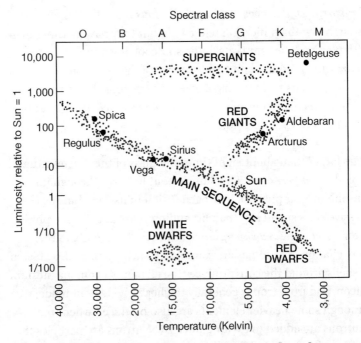

1. **The Hertzsprung–Russell diagram for the Sun and nearby stars. Temperatures are in Kelvin, which is 273 plus the temperature in degrees Celsius; the Kelvin scale is defined so that 0 K or 'absolute zero' is where all molecular movement stops**

forms the basis of new generations of stars and abodes for life, their planets. In the case of stars reaching tens of solar masses, although a supernova still occurs, the collapse at the centre produces a *black hole*—an object so massive that nothing escapes its gravity, including light.

The phase when a star converts hydrogen to helium in its core is the *main sequence* lifetime, which is generally the interval we consider optimal for life to thrive on planets around the star. The 'main sequence' refers to a diagonal swathe from upper left to lower right on the most famous graph in astronomy, the *Hertzsprung–Russell (H–R) diagram* (Fig. 1), named after its two originators. The graph plots a star's luminosity versus its surface temperature. By a 'surface', astronomers don't mean a hard surface but the level in a star's atmosphere where most light emerges, which is as deep as we can see.

Oddly, the temperature axis runs backwards from high to low in the H–R diagram. The purpose is to match the colour coding of stars from hot blue-white stars to cooler red ones with the letters O, B, A, F, G, K, and M. The letter gives a star's *spectral class*. Generations of astronomy students have remembered spectral types with the mnemonic 'Oh Be A Fine Girl/Guy Kiss Me!' (the letters don't stand for anything and have origins in 19th-century astronomy, which need not concern us).

On the main sequence, the most massive stars plot at the upper left and the lightest at the lower right. Whatever its mass, a main sequence star is called a dwarf—such as the Sun, a G-type dwarf. The lifetime on the main sequence can be over 50 billion years for cool red dwarfs.

The importance to astrobiology of the way stars 'live' and 'die' is wide ranging. Our Sun is a middle-aged, main sequence star with stable sunlight that fuels most life on Earth through photosynthesis. As mentioned at the outset, the atoms of life were generated in red giants and supergiants. Also, oxygen, silicon, magnesium, and iron are made from integral numbers of helium nuclei and so are particularly abundant products of nuclear fusion, which is significant because these atoms are the ones that make rocks. Rocky planets, like the one we inhabit, are a natural consequence of the physics of starlight. We also know that the Sun is at least a second-generation star because we have supernova elements on Earth such as gold. But since the Sun is only 4.6 billion years old in a 13.2 billion-year-old Milky Way, many stars came and went before life arose on Earth. Did earlier stars support planets and life, or even intelligent life, and what happened to them? This leads us to the question of how our own planet formed.

Getting a place to live: where planets come from

Ideas for the origin of the Solar System have a long heritage. In 1755, Immanuel Kant suggested that the Solar System coalesced out of a diffuse cloud in space. Later, in 1796, the mathematician Pierre Simon, the Marquis de Laplace, elaborated the notion. The basic Kant–Laplace concept is known as the 'nebular hypothesis,' after *nebula*, the Greek word for cloud.

The hypothesis starts with the idea that some part of the cloud is slightly denser than others and attracts material by gravity. It is likely that the cloud also has some slight initial rotation, so that the shrinking cloud spins faster, like an ice skater drawing in her arms. The random motion of the gas and dust will oppose material attracted along the axis of rotation but matter converging in the plane of rotation will also be resisted by the spin. As a result, the cloud flattens into a disc. The Sun forms in the centre, while planets coalesce in the plane of the disc from sparse material. This is consistent with the mass of the planets being only 0.1 per cent of the Sun's mass. Because they form from a disc, planets will all be in the same plane and they will all orbit the Sun in the same direction, as we observe.

In recent decades, it was thought that evidence from isotopes suggested that a shock wave from a nearby supernova triggered the collapse of the nebula. Isotopes are atoms that contain the same number of protons in their nucleus but a different number of neutrons. The Greek root, *isos topos*, means 'equal place', which refers to the same location in the periodic table of elements. Sometimes an isotope has a nucleus that is too big to be stable, and breaks apart by radioactive decay. An unstable aluminium-26 atom (which has a nucleus of 26 particles = 13 protons + 13 neutrons) decays into stable magnesium-26 (containing 12 protons + 14 neutrons). In any sample of aluminium-26 atoms, the time it takes for half of them to change into magnesium-26, the *half-life*, is 700,000 years. The presence of magnesium-26 in some meteorites, which must have been produced relatively quickly from aluminium-26, was thought to suggest a nearby supernova when the Solar System formed because massive stars make aluminium-26 and their supernovae distribute it.

However, in 2012, new measurements in meteorites showed that the levels of iron-60, which is an isotope only formed in supernovae, are too low for a nearby supernova. To explain abundant aluminum-26, a neighbouring massive star (perhaps more than twenty times the solar mass) of a type known as a Wolf–Rayet star may have shed its outer layers and spread aluminum-26 into the solar nebula. Aluminum-26 was a dominant heat source in the solar nebula for the first few million years. By melting ice in the earliest rocky material, the aluminum-26 caused water to go into hydrated minerals where it was safe, unlike ice that evaporates. If the aluminum-26 had not been present, the Earth might not have gained water-rich minerals and it might have neither oceans nor life.

The nebular hypothesis explains the broad distribution of different types of planet. The planets in the inner Solar System (Mercury, Venus, Earth, and Mars) are relatively small and rocky, while those in the outer Solar System (Jupiter, Saturn, Uranus, and Neptune) are giants. When the disc was forming, the matter drawn inwards gained energy of motion and the centre of the disc, where the Sun formed, became extremely hot. This created a temperature gradient from the hot centre of the disc to a cold exterior. Outside the Sun, pressures were low, which meant that substances existed either as solids or gases but not liquids. In the inner disc, it was far too hot for gases such as water vapour to form ice, but metal and rock vapour could condense nearly anywhere in the disc. Consequently, planets that formed in the inner region, such as the Earth, ended up with iron-rich cores surrounded by rocky mantles. In contrast, water condensed into ice from just inside the orbit of Jupiter ('the ice line') and farther out, providing more material to make larger planets. Methane could also condense as an ice from the orbit of Neptune outwards.

The giant planets are considered to have formed before the rocky ones. Jupiter and Saturn probably formed when rocky cores reached a size of about ten or so Earth masses that had sufficient gravity to attract more and more hydrogen and helium gas directly from the disc until all that was available in their orbit was sucked up. Because they are huge balls of mostly gas, we call these planets *gas giants*. This process happened within about 10 million years after the formation of the Sun. Uranus and Neptune are smaller, and drew in a greater proportion of icy solids, so we call these planets *ice giants*. Unlike the rapid growth of Jupiter, the formation of the inner rocky planets was spread over 100–200 million years. Planetesimals, which are 'pieces of planet', coalesced to make larger, rocky objects called planetary embryos with a size between that of the Moon and Mars. In turn, several planetary embryos merged into Venus and Earth and fewer into Mercury and Mars.

Although the inner planets accumulated material in their locality, gravitational nudges from the giant planets, particularly Jupiter, would have sent planetesimals careering into the inner Solar System. Scattered hydrated asteroids that originated from beyond the orbit of Mars were likely responsible for bringing water to the Earth that eventually turned into our oceans and lakes. We drink asteroid water. Computer simulations show that Jupiter ejected more water-rich material than it scattered inward, so that if Jupiter had had a less circular orbit and ejected even more water-rich material, Earth might have ended up without oceans and life.

We also know that planets do not necessarily stay put in their orbits. So-called *hot Jupiters* are exoplanets similar in mass to Jupiter that orbit at least twice as close to their host stars as the Earth orbits the Sun. Such exoplanets cannot have formed where they currently reside because it would have been too hot. It turns out that planets migrate because of large tails of gas and dust in the nebula that accompany their formation as well as the gravitational influence from other bodies. So the traditional nebula hypothesis is nuanced by *planetary migration* that could destroy or favour planetary habitability in extrasolar systems, depending on the details. In the next chapter, I'll discuss the idea that even the giant planets of our own Solar System may have migrated somewhat.

Nonetheless, the essential idea of the nebular hypothesis—that the planets formed from a disc—was confirmed in the 1980s when discs of debris around young stars were seen through telescopes. In fact, there is still some leftover debris in our own Solar System. Comets are icy bodies and asteroids are rocky rubble that were never assimilated. Occasionally, small chunks that have been knocked off asteroids through collisions end up falling onto the Earth's surface as meteorites. Consequently, meteorites provide us with key data about the early Solar System.

The age of the Earth and the Moon

Perhaps the most profound information gleaned from meteorites is the age of the Earth and Solar System. Ever since the 18th century, the vast layers of sedimentary rock seen by geologists had led them to suspect that the Earth, and hence the Solar System, must be of great age, but proof was lacking.

The first person to attempt to measure the age of the Earth was an English geologist, Arthur Holmes, who had the idea of examining lead isotopes formed from radioactive uranium. Uranium-238 decays into a cascade of further unstable isotopes of other elements until reaching a stable lead isotope, lead-206. The half-life for uranium-238 to change into lead-206 is 4.47 billion years. Another radioactive isotope, uranium-235, decays into lead-207 with a half-life of 704 million years. So, by measuring the amounts of lead-207 and lead-206 in different mineral grains of a rock, you can determine the age of the rock, because although there may have been different amounts of uranium in each grain, the fixed decay rates add the same ratio of lead-206 to lead-207 in a given time. In 1947, Holmes applied his method to a piece of lead ore from Greenland, and estimated that the Earth formed by 3.4 Ga (where Ga means 'Giga anna', equivalent to 1,000 million years or 'billions of years ago').

Holmes had two problems. First, he could never be sure that even the oldest rock he could find was as old as the Earth itself. Second, for a precise age, Holmes needed to know the small, original ratio of lead isotopes present when the Earth formed, so-called 'primeval lead', before subsequent uranium decay added further lead atoms. Clair Patterson, an American geochemist, realized that the first problem could be avoided by looking at meteorites because these are leftover building materials that formed around the same time as the Earth. He also recognized that certain types of meteorite, the iron meteorites, contain negligible uranium and so their ratio of lead isotopes provides a measure of primeval lead. With this approach, in 1953, Patterson accurately aged the Earth at 4.5 Ga. He was so excited that he feared a heart attack, and his mother had to take him to hospital.

Since then, improved techniques using radioactive isotopes have given us a timeline for the events surrounding the Earth's formation. The very oldest grains in meteorites suggest that the Solar System formed at 4.57 Ga. The Earth formed slightly later at 4.54 Ga. Then, around 4.5 Ga, the Earth was apparently hit by a Mars-sized object, which is named *Theia* after the Greek goddess who gave birth to the Moon goddess Selene. According to this *giant impact hypothesis*, debris that was blasted out from the impact went into orbit around the Earth and coalesced to form the Moon.

The Moon is important for astrobiology because its gravity stabilizes the tilt of Earth's axis to the plane of its orbit, which helps the Earth maintain a relatively steady climate. Currently, the Earth's axis is tilted 23.5 degrees, but if it were able to vary widely, great climatic swings would occur. For example, at 90 degrees tilt, the Earth would be tipped on its side and ice would form seasonally at the equator. Computer simulations show that if we had no Moon, the Earth's axial tilt would vary chaotically over periods of millions of years and have a large range, potentially from 0 to more than 50 degrees. Microbial life could probably withstand large climate swings. But advanced animal life and civilizations such as ours would be challenged.

In following our astronomical trail, we have now arrived at the point of the creation of an abode for life, our own planet. There were contingencies in whether the water necessary for life was delivered to the Earth and controls on where an Earth-like planet might occur. Then, after Earth formed, how and when did life arise?

Chapter 3
Origins of life and environment

The early Earth

So little is known about the very earliest aeon of Earth history that it doesn't even have an official name. Informally, it is called the Hadean, which started around 4.5 Ga when the Moon formed, and ended at a date that hasn't been agreed upon but is usually taken as either 3.8 or 4.0 Ga. It is probable that life originated in the Hadean, but so far we haven't found evidence because there are no sedimentary rocks from this time. Such rocks consist of layers of sediment grains laid down in water or from airfall, and so best preserve traces of biology or the environment. Consequently, our ideas of what happened during the Hadean have to be constructed from sparse data aided by the theoretical constraints.

Theory suggests that heat from the giant impact that formed the Moon would have turned rock into gas. The atmosphere of vaporized rock would have lasted for a few thousand years and then condensed and rained down on a molten magma surface, which eventually solidified into a crust. Subsequently, the atmosphere would have consisted mainly of extremely dense steam for a few million years before it condensed to form oceans.

When did continents begin to form? We get some clues from tiny mineral grains less than 0.5 mm across that are left behind from the first half billion years or so of Earth history. These are *zircons*, crystals of zirconium silicate with chemical formula $ZrSiO_4$, where Zr is zirconium, Si is silicon, and O is oxygen. Zircons are so tough that they remain even after the rock in which they were once hosted has eroded away. Some zircons as old as 4.4 to 4.0 Ga have been found in fossilized gravel in the Jack Hills, which are about a thousand kilometres north of Perth in western Australia. The zircons contain flecks of quartz, which is a crystalline form of silica, SiO_2. The quartz may have been derived from granites, which is the silica-rich type of igneous rock that makes up much of the continents. In this way, zircons suggest that Earth's continental crust existed as long ago as 4.3 Ga. Isotopes support this inference. Some ancient zircons are enriched in stable oxygen-18 atoms relative to stable oxygen-16 ones. Such enrichment occurs when surface waters make clays and the mud is then buried and melts underground, passing the isotopic signature to igneous rocks.

In the Hadean, the Earth was probably hit by a few huge pieces of debris left over from Solar System formation, but none as big as the Moon-forming impactor. The energy of a small number of very large impacts could have vaporized the entire ocean or its upper few hundred metres. If so, life would have had to restart or it might have been trimmed back to only those microbes sheltered underground that were able to survive the heat. In fact, partially sterilizing impacts may explain the nature of the last common ancestor for all life on Earth. Genetics traces the common ancestor to a thermophile (see Chapter 5)—a microbe that lives in hot environments. Essentially, DNA analysis implies that your 'great-great-great-...grandmother' was a thermophile, if you insert enough 'greats'. This may be because thermophiles were the only survivors of huge impacts.

The Earth today is not entirely free from potentially catastrophic impacts. For example, Chiron is a comet-like object about 230 km across in the outer Solar System that crosses the orbit of

Saturn. One day within the next 10 million years or so, a nudge from Saturn's gravity will fling Chiron towards the Sun or away from it. In the former case, Chiron's chances of impacting the Earth would be less than one in a million. But if Chiron did hit, the heat would turn the upper few hundred metres of the ocean into steam, and land would be sterilized down to about fifty metres' depth. Thermophiles might become the ancestors of subsequent life in a sort of evolutionary déjà vu, if the impact explanation for a thermophile ancestor is correct.

The last hurrah of the big Hadean impacts is known as the *Late Heavy Bombardment*, which happened about 4 to 3.8 Ga. Craters on the Moon bear witness to this massive bombardment. Rocks brought back by Apollo astronauts have been dated using radioisotopes and indicate that many craters were produced within the same 200-million-year interval.

The leading hypothesis for explaining the Late Heavy Bombardment was devised by astronomers in Nice, France, and is therefore called the *Nice model*. It relies on the astonishing idea that the orbits of Jupiter and the other giant planets shifted at the end of the Hadean. Calculations suggest that after the Solar System formed, mutual gravitational effects caused Saturn and Jupiter to reach a state called a 'resonance' in which Saturn orbited the Sun once for every two Jupiter orbits. Regular alignment created periodic gravitational prodding that made the orbits of Saturn and Jupiter less circular. In turn, the orbits of Neptune and Uranus were perturbed and moved outwards, also becoming more elliptical. Neptune could have even begun inside the orbit of Uranus and then overtaken it in a migration outward, and both planets, particularly the farther one, would have scattered small icy bodies, some towards the inner Solar System. Meanwhile, the gravity and movement of Jupiter would fling some asteroids into the inner Solar System and push others away. During this time, the Earth must have suffered even more impacts than the Moon because of Earth's greater size and gravity.

Eventually, after wreaking havoc, the orbits of the giant planets would have settled down.

The origin of life

Quite how life arose is unknown. It may have originated on Earth or it was carried here by space dust or meteorites. The latter idea is called *panspermia*, and doesn't solve how life originated, but pushes the problem elsewhere. Also, there may be difficulties with survival over long transit times if life came from around other stars. For these reasons, we'll concentrate on a terrestrial origin of life.

There is wide agreement that the origin of life would have been preceded by a period of chemical evolution, or *prebiotic chemistry*, during which more complex organic molecules were produced from simpler ones. The idea goes back to the 19th century. In 1871, Charles Darwin imagined that such chemistry might have occurred in a 'warm little pond', according to a letter to the botanist Joseph Hooker:

> It is often said that all the conditions for the first production of a living organism are now present, which could ever have been present.—But if (and oh what a big if) we could conceive in some warm little pond with all sorts of ammonia and phosphoric salts,—light, heat, electricity etc. present, that a protein compound was chemically formed, ready to undergo still more complex changes, at the present day such matter would be instantly devoured, or absorbed, which would not have been the case before living creatures were formed.

Thus, Darwin concluded that life is unlikely to originate today because organisms are continually eating the chemical compounds that are needed. On the other hand, before life existed, chemical conditions for the origin of life would have been prevalent.

In the 1920s, the Russian biochemist Alexander Oparin and the British biologist J. B. S. Haldane both recognized that the environment under which life arose would have lacked oxygen. Earth's oxygen-rich atmosphere is a product of photosynthesis by plants, algae, and bacteria. An oxygen-free atmosphere would have been better suited to prebiotic chemistry because oxygen converts organic matter into carbon dioxide, which prevents the build-up of complex molecules. Indeed, a contemporary geologist, Alexander MacGregor, reported in 1927 that he had found sedimentary rocks dating from within the Archaean Aeon (3.8–2.5 Ga) that showed that the ancient atmosphere lacked oxygen. In particular, MacGregor observed that iron minerals in what is now Zimbabwe were unoxidized, unlike the rust-coloured iron oxides in modern sediments that are produced when atmospheric oxygen reacts with iron-containing minerals. Today, MacGregor's deduction is supported by many other data (Chapter 4).

Oparin and Haldane also proposed that gases in Earth's early atmosphere would have been converted by ultraviolet sunlight or lightning into organic molecules. In the 1950s, such ideas were tested when Stanley Miller, a student of the Nobel Prize-winning chemist Harold Urey, devised an experiment at the University of Chicago. A blend of ammonia, methane, hydrogen, and water vapour was put in a flask and electric sparks passed through the gas to simulate lightning. Miller found that yellowish water condensed out at the bottom of the flask. This dark residue contained organic molecules, including various amino acids—the building blocks of proteins. At the time, the Miller–Urey experiment was trumpeted in the media as virtually solving the problem of the origin of life. Life was said to start in a 'primordial soup' generated by atmospheric chemistry. In 1953, when Miller wrote up his results, a common view (dating back to Oparin and Haldane) was that genetic material was protein, but a month before Miller's paper came out, DNA was identified as the real basis of heredity. Subsequently, experiments in the Miller–Urey genre have also produced molecules containing hexagonal rings of carbon, nitrogen, and hydrogen atoms, which are the kind of rings found in DNA.

However, geochemists have raised doubts about the Miller–Urey experiment because they argue that the Earth's early atmosphere was probably not as hydrogen rich as Miller assumed. Volcanoes supply the Earth's atmosphere with gases over long timescales but most volcanic gas is steam, i.e. water vapour (H_2O), with an average of less than 1 per cent hydrogen (H_2). Similarly, volcanic carbon comes out as carbon dioxide (CO_2) rather than the hydrogenated form of methane (CH_4), while nitrogen is emitted as dinitrogen (N_2) instead of ammonia (NH_3).

It is thought that the atmosphere in the Hadean was mostly maintained from gases released through volcanism, so the air should have consisted predominantly of CO_2 and N_2. Of course, it depends on how far back you go. When the Earth was forming, the vaporization of large impactors would have produced hydrogen-rich atmospheres if impactors were sufficiently rich in iron to provide the right kind of chemistry to stabilize hydrogen in the impact explosion. Early on, most of the atmosphere was steam and oceans had not yet condensed. After oceans formed, atmospheres produced by impact vaporization would have been ephemeral. Because of uncertainties about the composition of the Hadean atmosphere, astrobiologists have proposed sources of organic carbon other than that presumed by Miller and Urey.

One alternative is that organic carbon came from space. Some carbon-rich meteorites contain amino acids, alcohols, and other organic compounds that perhaps could have seeded Earth with the materials needed for prebiotic chemistry. For example, the Murchison meteorite, which fell on Murchison, Australia, in 1969, contains over 14,000 different molecules. Also, micrometeorite particles that are 0.02–0.4 mm in size currently bring about 30 million kg of organic carbon to the Earth per year. Organic carbon can fall intact to the Earth's surface because tiny particles don't necessarily burn up in the atmosphere. Furthermore, the interiors of comets are rich in organic material and may have played a similar role to meteorites in getting life started if the organic material survived cometary impacts on the Earth.

Another possible source of organic carbon is deep-sea hydrothermal vents. A hydrothermal vent is a spot where hot water emerges from the seafloor. New seafloor is created at mid-ocean ridges, where the Earth's tectonic plates separate and magma rises up from the mantle, filling the void. Seawater flows down through cracks near the ridges, is heated by the hot magma, and rises, picking up substances produced by chemical reactions between the water and rocks, such as hydrogen, hydrogen sulphide, and dissolved iron. When the hot acidic water hits cold (2°C) ocean water, it produces a plume of precipitated particles containing dark-coloured pyrite, an iron sulphide mineral (FeS_2, where Fe is iron and S is sulphur). That is why these hydrothermal vents are called 'black smokers'. One hypothesis suggests that life originated on the surface of such pyrite minerals. However, black smokers are very hot, around 350°C, so most researchers who favour a deep-sea origin of life have focused on vents that are cooler, around 90°C, alkaline rather than acidic, and located further from the mid-ocean ridges.

Beneath alkaline vents, hydrogen released in reactions between water and rock can combine with carbon dioxide to make methane and bigger organic molecules. In the Hadean, tiny pores in the mineral structures that grew up from vents could, in principle, have contained and concentrated simple molecules to allow them to react and make more complex prebiotic molecules.

Also, in alkaline vents, because the fluid emerging is more alkaline than seawater, there is a natural gradient in pH that is similar to that in cells. If life originated in such an environment, it might explain why energy production is intimately tied to pH gradients.

Life generates energy from microscopic electrical motors that are embedded in cell membranes and run off electrical currents driven by pH gradients across the membranes. It is impossible for words to do justice to these amazing molecular machines, so I suggest that the reader search the Internet for 'ATP synthase animations'. Essentially, metabolic energy is used to create a different pH on each side of a cell membrane with the outside usually more acidic than the inside. This differential is a *proton gradient*, because pH is an inverse measure of the concentration of positively charged hydrogen ions (protons) in solution—the greater the concentration, the lower (and more acidic) the pH. The proton gradient is essentially a battery. When discharged, electrical current flows through the molecular turbine in the cell membrane and generates molecules that store energy. These molecules of adenosine triphosphate or 'ATP' are the energy carriers in all cells. For example, a human typically contains 250g of ATP and uses up a body weight's worth of ATP every day as ATP is made used, and remade.

The above process of generating ATP is called *chemiosmosis*, which contains the word 'osmosis', meaning water moving across a membrane, because the proton movement is analogous. Chemiosmosis is so weird compared with the familiar idea of generating energy from chemical reactions that when it was proposed in 1961 by British biochemist Peter Mitchell most biochemists were hostile. Eventually, Mitchell was vindicated with a Nobel Prize in 1978. Given that chemiosmosis is universal to terrestrial life and presumably related to its origin, it is interesting to wonder whether ion gradients might be fundamental for life anywhere.

Currently, the source of the organic carbon that led to the origin of life—the atmosphere, space, or hydrothermal vents—is unresolved. Given such a source, however, organic compounds must have been concentrated for chemical reactions to happen, which could have occurred when films formed on mineral surfaces, perhaps arising from drying or coldness. Then the simple organics could join up to make more complex molecules. But chemical systems need several other features before they can be considered 'alive'. The main ones are a metabolism, enclosure by a semi-permeable membrane, and reproduction that incorporates heredity with a genome.

With regard to the last characteristic, a more tractable issue for laboratory experimentation than the origin of life itself is a hypothesized stage of very early life, called the *RNA World*, in which RNA preceded DNA as the genetic material. In modern cells, DNA (deoxyribonucleic acid) provides an instruction set that is transcribed into RNA (ribonucleic acid) molecules that carry the information needed to make specific proteins. DNA is similar to RNA but more complicated, so it is logical that RNA evolved first. Moreover, in the 1980s it was discovered that some RNA molecules, known as ribozymes, can act as catalysts. So it is plausible that RNA once catalyzed its self-replication and assembly from smaller molecules. At the next stage, RNA would begin to make proteins, some of which would be better catalysts than RNA itself. Eventually, DNA would replace RNA because it is more stable and can be larger, both of which confer reproductive advantages.

Another critical feature in the origin of life was the encapsulation of a genome in a cell membrane. This was beneficial for a couple of reasons: first, to increase rates of reaction by concentrating biochemicals; and second, to give an evolutionary advantage to self-replicating molecules. For example, an RNA genome that produced a useful protein could keep it to itself inside a cell, thereby favouring its progeny.

So-called 'pre-cell' structures have been made in the lab. Spherical structures form spontaneously when oils are mixed with water, which is something we're all familiar with in the kitchen. Such oily or fatty membranes plausibly led to the membranes that surround modern cells and were surfaces for prebiotic chemistry.

Subsequent events may have occurred as follows. RNA took up residence in pre-cells, cells with an RNA genome evolved, and then modern cells with a DNA genome took over from RNA. At some point, metabolism also evolved, perhaps after RNA World. However, there remains an ongoing debate about which came first, metabolism or genetic replication. Also, the actual location, conditions, and evolutionary steps for the origin of life remain uncertain and are an area of research ripe for significant progress.

Chirality (or the art of clapping with one hand)

A description of life's origin would be incomplete without mentioning a related aspect of biochemistry: for biomolecules that come with two mirror-image structures, life on Earth uses only one of the structures, not both. The property of having mirror-image symmetry is called *chirality*, which comes from the Greek *cheir* for 'hand', given the fact that a person's hands are mirror images and cannot fit on top of one another (the same way up). If you're not convinced that a left-hand structure is truly different from a right one, try wearing your shoes on the wrong feet all day!

Chirality arises in biomolecules if a central carbon atom is surrounded by four different groups of atoms. The groups naturally arrange themselves in a tetrahedron, which is a pyramid with four triangular faces. If we number the groups with a '1' at the top of the tetrahedron, the three tetrahedral feet can be numbered clockwise or counterclockwise as shown:

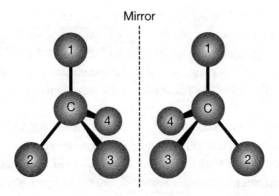

The two molecules in the picture are 'handed' because one cannot be superimposed on top of the other. Such molecules are called *enantiomers* from the Greek *enantios morphe*, for opposite shape. Let's consider a specific enantiomer, the amino acid alanine. It would have '1' as a hydrogen atom, '2' as a carboxyl group (COOH), '3' as an amine group (NH_2), and '4' as a methyl group (CH_3). By convention, with the clockwise numbering of groups 2 to 4, alanine is said to be in the right-handed or D-form after the Greek *dextro* for right. With the counterclockwise arrangement, alanine is in the L form from the Greek *laevo* for left. Remarkably, life on Earth uses L-amino acids in proteins, almost entirely, and D-sugars almost exclusively as well whether in DNA, cell walls, or synthesis. (A handy mnemonic is 'lads' for left amino, dextro sugars). However, when a chiral substance is made in the laboratory, equal proportions of left- and right-handed forms are usually produced, called a *racemic* mixture.

Chiral molecules can have very different effects despite their identical chemical formulae. An example is the infamous drug thalidomide. The right-handed form cures morning sickness in pregnant women, but the left-handed version induces severe birth defects. A more pleasant example is the flavouring limonene: the D-form tastes like lemons but the L-form is orange-flavoured.

What caused the common chirality, or *homochirality*, shared by all life? There are two general hypotheses. One is that the organic molecules from which life started had a slight excess of one enantiomer that was subsequently amplified. Physical processes invoked to induce this excess include polarized light from stars that might cause organic molecules that seeded the Earth to develop a chiral bias during their synthesis in space; alternatively, polarized radiation from radioactive decay might generate an enantiomeric excess. A problem with such radiation-related ideas is that the excess is often small, less than 1 per cent. The second hypothesis is that prebiotic chemistry preferentially assembled only one of the enantiomers. For example, adsorption on a mineral surface might have chiral selectivity. Chemical kinetics is generally more efficient for a pure substance than a mixture, so this might cause further amplification. One biological process—reproduction of a genome—may require homochirality. Homochirality was probably needed for the replication in the RNA World model because only replicated RNA strands of the same chirality would line up with an RNA template strand.

Signs of the earliest life

Although the origin of life remains obscure, in the oldest sedimentary rocks on Earth we find possible signs of life. Perhaps it's no coincidence, but these rocks (mixed with volcanic flows) date from the end of the Late Heavy Bombardment, around 3.8 Ga. They are found in Isua, in the interior of south-west Greenland. They form a dark, foreboding landscape that looks something like Mordor, albeit icy, and they are remarkable for telling us much about the early Earth. There is gravel that was rounded by the action of water. Also, some sediments that were laid down under the sea contain flakes of carbon in the form of graphite. Isotopic analysis suggests that this carbon was once part of microbes living in the oceans. Carbon has two stable isotopes, carbon-12 and carbon-13. Life tends to concentrate a few per cent more carbon-12 atoms in its tissues than carbon-13 because molecules containing lighter carbon-12 atoms react faster in biochemistry. The graphite in Isua is enriched in carbon-12 by about 2 per cent, which is similar to the proportion in marine microbes. So the inference is that marine microbes died and sank into the sediments where the carbon was subsequently compressed and converted into graphite.

Better preserved evidence for early life is found in north-west Australia and dates from about 3.5 Ga. In the 1980s, my colleague Roger Buick found fossil stromatolites of this age near North Pole in north-western Australia (Fig. 2). An Aussie joker named the place 'North Pole' because it is one of the hottest, most sun-baked locations on Earth. Stromatolites are laminated sedimentary structures made by photosynthesizing microbial communities in water that is shallow enough to receive sunlight. Often stromatolites take the form of domes of wrinkly laminations. These domes are built up from trapped sediment as sheets of microbes, called *microbial mats*, growing upwards towards sunlight. On modern stromatolites, the living part—the mat—is just a veneer about a centimetre thick with a texture like tofu that sits atop accumulated

2. *Left*: Cross-section of the world's oldest fossil stromatolites in the Dresser Formation, North Pole, Australia. *Right*: A plan view of the bedding plane of the stromatolites, showing their tops. The lens cap is 5 cm diameter for scale

mineral layers. If you could travel back to the Archaean Earth, you would see coastlines covered in stromatolites all over the world. Today, stromatolites are rare because certain fish and snails (which didn't exist in the Archaean) feed on the microbial mats that make them. Consequently, modern stromatolites are found only in lagoons or lakes where the water is too salty for such voracious animals.

Some sceptics wonder if the North Pole structures could have formed without life and so are not really stromatolites, but most astrobiologists accept their organic heritage because they have various features we expect of biology. The fossil stromatolites have laminations with extremely irregular wrinkles that are difficult to explain as folds created by purely physical processes. Also, the laminations thicken at the top of convex flexures, which is just what happens when photosynthetic microbes grow where there is more sunlight. In troughs between flexures there are fragments that are plausibly interpreted as pieces of microbial mat that were ripped up by waves in the sea. The fragments contain thin, wrinkly layers that incorporate organic carbon. The combination of such observations suggests that the North Pole stromatolites are indeed the oldest fossil structures visible to the naked eye.

To find evidence for past life, we can also look for individual dead bodies of single-celled organisms, since only microbes existed so far back in Earth history. Such *microfossils*, which can only be seen with a microscope, are found in sedimentary rocks such as chert (a fine-grained silica-rich rock; flint is a familiar example) and shales, which are fine-grained sedimentary rocks that were once muds.

Geologists who study microfossils take rock samples from likely settings back to the lab where they slice them up and hope that something shows up under the microscope. It's a hit-and-miss affair. The problem of distinguishing life from non-life also rears its ugly head again because mineral grains sometimes look like fossil cells. Consequently, microfossils are more convincing if they show signs of cell division, colonial behaviour, or filamentous structures that occur when cells join up in a line. Microfossils can also be tested to see if they contain organic carbon.

The oldest incontestable microfossils occur in South Africa from 2.55 Ga. They are convincing because they are found in fossilized mats of filaments that contain organic carbon and include some microbial forms immobilized in the act of cell division. There are older possible microfossils in South Africa dated at 3.2–3.5 Ga that are spheroidal, contain organic carbon, and appear to show cell division, but have no other biological attributes. Similar microfossil colonies in 3.4 Ga sandstone of the Strelley Pool Formation in north-west Australia also possess organic carbon with isotopes characteristic of life.

The world's most controversial microfossils come from the Apex Chert rock formation in north-west Australia, with an age of 3.5 Ga. These were identified as the 'world's oldest microfossils' in the 1990s by Bill Schopf, of the University of California, Los Angeles. The structures appear as black to dark brown streaks that appear to be partitioned into cell-like sections. Schopf gave them Latin names implying that they were extinct cyanobacteria, which are photosynthetic bacteria. In 2002, a rancorous dispute arose when Martin Brasier of Oxford University re-examined the samples and showed that they didn't look like microbial filaments in three dimensions. Brasier also noted that the chert was not sedimentary but a type produced in a hydrothermal vent, which is an unlikely place to find cyanobacteria that need sunlight. Since then, further investigation has suggested that the filaments are really fractures that are partially filled with haematite, an iron oxide, which provides the dark colour. However, organic carbon is dispersed in the chert outside the filaments and the controversy continues.

So far, we've discussed isotopes, microfossils, and stromatolites. Further possible indicators of early life are *biomarkers*, which are recognizable derivatives of biological molecules. We're all familiar with the concept that fossil skeletons allow us to distinguish a *Tyrannosaurus rex* or *Triceratops*. On the microscopic scale, microbes can leave behind remnants of individual organic molecules, consisting of skeletal rings or chains of carbon atoms. These 'molecular backbones' come from particular molecules that are found only in certain types of microbe. So biomarkers not only confirm the presence of past life but can also indicate specific forms of life.

Unfortunately, like some microfossils, ancient biomarkers are also mired in controversy. The oldest reported biomarkers date from about 2.8 Ga and appear to show molecules suggesting the presence of oxygen-producing cyanobacteria. But the sampled rocks may have been contaminated by younger organic material. More certain biomarkers are found in rocks from 2.5 Ga inside tiny fluid inclusions that appear uncontaminated.

Despite the limitations of some evidence, the overall record shows that Earth was inhabited by 3.5 Ga—within 1 billion years of its formation and shortly after heavy impact bombardment, which suggests that life might originate fairly quickly on geological timescales on suitable planets. This means that it's not out of the question that Mars, which we think had a geologically short window of habitability (Chapter 6), may also have evolved life.

Chapter 4
From slime to the sublime

How did Earth maintain an environment fit for life?

However life started, once established, it persisted for over 3.5 billion years and evolved from microbial slime to the sophistication of human civilization. During this period, the Earth maintained oceans and, for the most part, a moderate climate, even though there was an increase of about 25 per cent in the amount of sunlight. The gradual brightening of the Sun is a consequence of the way that the Sun shines on the main sequence. When four hydrogen nuclei are fused into one helium nucleus in the Sun's core, there are fewer particles, so material above the core presses inwards to fill available space. The compressed core warms up, causing fusion reactions to proceed faster, so that the Sun brightens about 7–9 per cent every billion years. This theory is confirmed by observations of Sun-like stars of different ages.

If the Earth had possessed today's atmosphere at about 2 Ga or earlier, the whole planet should have been frozen under the fainter Sun, but geological evidence suggests otherwise. This puzzle is the *faint young Sun paradox*. There is evidence for liquid water back to at least 3.8 Ga, which includes, for example, the presence of sedimentary rocks that were formed when water washed material from the continents into the oceans.

There are three ways to resolve the faint young Sun paradox. The most likely explanation involves a greater greenhouse effect in the past. Another suggestion is that the ancient Earth as a whole was darker than today and absorbed more sunlight, although there's scant supporting evidence. A third idea is that the young Sun shed lots of material in a rapid outflow (solar wind) to space so that the Sun's core wasn't as compressed and heated over time by overlying weight as assumed above. If the mass loss was just right, the Sun could have started out as bright as today. However, observations of young Sun-like stars presently don't support the third hypothesis.

In considering the first idea, we need to appreciate that the atmosphere warms the Earth to an extent depending on atmospheric composition. Without an atmosphere (and assuming that the amount of sunlight that the Earth reflects stayed the same), the Earth's surface would be a chilly −18°C. Instead, today's average global surface temperature is +15°C. The 33°C difference (= 18 + 15) is the size of Earth's *greenhouse effect*, which is the warming caused by the atmosphere.

How does the greenhouse effect work? A planet's surface is heated by visible sunlight, causing it to glow in the infrared just as your warm body shines at those wavelengths. An atmosphere tends to be mostly opaque to the infrared radiation coming up from the surface below, so it absorbs the infrared energy and warms. Because the atmosphere is warm it also radiates in the infrared. Some of this radiation from the atmosphere travels back down to the planet. So the surface of a planet is warmer than it would be in the absence of an atmosphere because it receives energy from a heated atmosphere in addition to visible sunlight.

At 3.5 Ga, when the Sun was 25 per cent fainter, a 50°C greenhouse effect would have been needed to maintain the same global average temperature on Earth that we enjoy today with a 33°C greenhouse effect. A stronger greenhouse effect is possible if there had been greater levels of greenhouse gases, which are those responsible for absorbing infrared radiation coming from the Earth's surface. Today, water vapour accounts for about two-thirds of the 33°C greenhouse effect, and carbon dioxide (CO_2) accounts for most of the rest. However, water vapour condenses as rain or snow, so its concentration is basically a response to the background temperature set by atmospheric CO_2, which doesn't condense. In this way, CO_2 controls today's greenhouse effect even though its level is small. Around 1700, there were about 280 parts per million of CO_2 in Earth's atmosphere (meaning that in a million molecules of air, 280 were CO_2), while in 2010 there were 390 parts per million. Since industrialization, CO_2 has been released from deforestation and burning fossil fuels such as oil or coal. In the late Archaean, an upper limit on the CO_2 amount of tens to a hundred times higher than the pre-industrial level is deduced from the chemical analyses of *palaeosols*, which are fossilized soils. But even such levels of CO_2 were not enough to counter the faint young Sun.

In fact, the Archaean atmosphere had less than one part per million of molecular oxygen (O_2), which implies that methane (CH_4) was an important greenhouse gas. Today, atmospheric methane is at a low level of 1.8 parts per million because it reacts with oxygen, which is the second most abundant gas in the air at 21 per cent. (Most of today's air is nitrogen (N_2), which is 78 per cent, but neither oxygen nor nitrogen are greenhouse gases.) In the Archaean, the lack of oxygen would have allowed atmospheric methane to reach a level of thousands of parts per million. Methane wouldn't accumulate without limit because it can be decomposed by ultraviolet light in the upper atmosphere. Subsequent chemistry involving methane's decomposition fragments can generate other hydrocarbons, i.e. chemicals made of hydrogen and carbon, including ethane gas (C_2H_6) and a smog of sooty particles. A combination of methane, ethane, and carbon dioxide—plus the water vapour that builds up in response to the temperature set by these non-condensable greenhouse gases—would have provided enough greenhouse effect to offset the fainter Sun. This assumes that there was a source of methane from microbial life, just like today, which is plausible because the metabolism for methane generation is ancient (see Chapter 5).

Of course, we could ask what the Earth's early climate was like before life. In that case, CO_2 probably controlled the greenhouse effect, as today. In fact, throughout much of Earth history, there has been a geological cycle of CO_2 that regulated climate on timescales of about a million years. (It still operates today but is far too slow to counteract human-induced global warming.) Essentially, atmospheric CO_2 dissolves in rainwater and reacts with silicate rocks on the continents. The dissolved carbon from this *chemical weathering* then travels down rivers to the oceans, where it ends up in rocks on the seafloor, such as in limestone, which is calcium carbonate ($CaCO_3$). If deposition of carbonates were all that were happening, the Earth would lose atmospheric CO_2 and freeze, but there is a mechanism that returns CO_2 to the atmosphere. Seafloor carbonates are transported on slowly moving oceanic plates that descend beneath other plates in the process of *subduction*. An example today is the South Pacific 'Nazca' plate, which is sliding eastwards under Chile. Carbonates are squeezed and heated during subduction, causing them to decompose into CO_2. Volcanism (where rocks melt) and metamorphism (where rocks are heated and pressurized but don't melt) release CO_2. The whole cycle of CO_2 loss and replenishment is called the *carbonate–silicate cycle* and behaves like a thermostat. If the climate gets warm, more rainfall and faster weathering consume CO_2 and cool the Earth. If the Earth becomes cold, CO_2 removal from the dry air is slow, so CO_2 accumulates from geological emissions, increasing the greenhouse effect.

The carbonate–silicate cycle probably regulated climate before life originated. It then likely played an increasingly important role as a thermostat after levels of methane greenhouse gas declined in two steps when atmospheric oxygen concentrations increased, first around 2.4 Ga and then 750–580 Ma (Ma = millions of years ago).

A caveat is that the cycle may have operated differently in the Hadean and possibly the Archaean because *plate tectonics*—the large-scale motion of geologic plates that ride on convection cells within the mantle below—probably had a different style. Radioactive elements in the mantle generate heat when they decay, and more were decaying on the early Earth. So, on the one hand, a hotter, less rigid Hadean mantle should have allowed oceanic crust to sink more quickly. On the other hand, a hotter mantle should have produced more melting and thicker, warmer oceanic crust that was less prone to subduct. Overall, the presence of granites inferred from zircons (Chapter 3) implies that crust must have been buried somehow because granite is produced when sunken crust melts. However, exactly how tectonics operated on the early Earth remains an open question.

The Great Oxidation Event: a step towards complex life

The most drastic changes of the Earth's atmospheric composition have been increases in oxygen, which were just as important for the evolution of life as variations in greenhouse gases. For most of Earth's history, oxygen levels were so low that oxygen-breathing animals were impossible. The *Great Oxidation Event* is when the atmosphere first became oxygenated, 2.4–2.3 Ga. However, oxygen levels only reached somewhere between 0.2 to 2 per cent by volume, not today's 21 per cent. Large animals were precluded until around 580 Ma after oxygen had increased a second time to levels exceeding 3 per cent (Fig. 3).

Nearly all atmospheric oxygen (O_2) is biological. A tiny amount is produced without life when ultraviolet sunlight breaks up water vapour molecules (H_2O) in the upper atmosphere, causing them to release hydrogen. Net oxygen is left behind when the hydrogen escapes into space, thereby preventing water from being reconstituted. But abiotic oxygen production is small because the upper atmosphere is dry. Instead, the major source

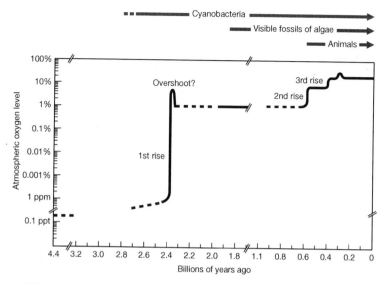

3. The approximate history of atmospheric oxygen, based on geologic evidence (ppm = parts per million; ppt = parts per trillion)

of oxygen is *oxygenic photosynthesis*, in which green plants, algae, and cyanobacteria use sunlight to split water into hydrogen and oxygen. These organisms combine the hydrogen with carbon dioxide to make organic matter, and they release O_2 as waste.

Another, more primitive, type of microbial photosynthesis that doesn't split water or release oxygen is *anoxygenic photosynthesis*. In this case, biomass is made using sunlight and hydrogen, hydrogen sulphide, or dissolved iron in hydrothermal areas around volcanoes. Today, microbial scum grows this way in hot springs.

Before plants and algae evolved, the earliest oxygen-producing organisms were similar to modern cyanobacteria. Cyanobacteria are bluish-green bacteria that teem in today's oceans and lakes. DNA studies show that a cyanobacteria-like microbe was the ancestor of plants, algae, and modern cyanobacteria. Consequently, we might suppose that the atmosphere became oxygenated once cyanobacteria evolved. However, evidence suggests that cyanobacteria were producing oxygen long before it flooded the atmosphere. A plausible explanation is that *reductants*, which are chemicals that consume oxygen, at first rapidly overwhelmed the oxygen. Reductants include gases such as hydrogen, carbon monoxide, and methane that come from volcanoes, geothermal areas, and seafloor vents. Distinctive iron-rich sedimentary rocks called *banded iron formations* dating from the Archaean show that there was considerable dissolved iron in the Archaean ocean, unlike today's ocean, which would have also reacted with oxygen.

The first signs of photosythetic oxygen appear about 2.7–2.8 Ga according to evidence from stromatolites and the presence of chemicals that became soluble by reacting with oxygen. In north-west Australia, stromatolites in rocks called the Tumbiana Formation once ringed ancient lakes. In theory, microbes might have built such stromatolites using anoxygenic photosynthesis, but there's no evidence for hydrothermal emissions needed for this metabolism. Instead, cyanobacteria using oxygenic photosynthesis probably constructed the stromatolites, consistent with tufts and pinnacles in the stromatolites that are produced when filaments of cyanobacteria glide towards sunlight. Furthermore, molybdenum and sulphur, which are elements only soluble when oxidized, become concentrated in seafloor sediments after 2.8 Ga to levels that are possible if microbes oxidized these elements on land using local sources of oxygen such as stromatolites.

Another reason why oxygen didn't immediately accumulate in the Earth's atmosphere is that its production is mostly a zero-sum process. When a molecule of oxygen is made, an accompanying molecule of organic carbon is generated, as summarized in the following equation:

carborn dioxide + water + sunlight = organic matter + oxygen

$$CO_2 + H_2O + \text{sunlight} = CH_2O + O_2$$

The process easily reverses, i.e. organic matter reacts with oxygen to regenerate carbon dioxide and water. Also, the Archaean atmosphere's lack of oxygen meant that microbes could readily convert organic carbon into methane gas, as happens today in smelly, oxygen-free lake or seafloor sediments. In the air, methane could react with the oxygen to recreate water vapour and carbon dioxide. In both cases—direct reversal or indirect cancellation with methane—no net oxygen is produced, despite the presence of photosynthesis.

However, a tiny fraction (about 0.2 per cent, today) of organic carbon is buried in sediments and separated from oxygen, which prevents the two from recombining. Every organic carbon that gets buried is equivalent to one net O_2 molecule that's freed up. Of course, this 'freed' oxygen can easily react with many other substances besides organic carbon, including geological gases and dissolved minerals such as iron. Nonetheless, a tipping point was reached when the flow of reductants to the atmosphere dropped below the net flow of oxygen associated with organic carbon burial, causing the Great Oxidation Event. Afterwards, a plateau of 0.2–2 per cent oxygen was reached probably because oxygen that dissolved in rainwater reacted significantly with continental minerals, preventing its further accrual.

Indeed, a surge in oxidation is the evidence for the Great Oxidation. 'Red beds'—rust-coloured continental surfaces that arise when iron minerals react with oxygen—appear. Today, reddish land surfaces are common, such as in the American south-west. A change in sulphur processing in the atmosphere also occurred. Before the Great Oxidation, when there was less than one part per million of atmospheric oxygen, red- and yellow-coloured particles of elemental sulphur literally fell out of the sky. They formed in chemical reactions at many kilometres altitude when sulphur dioxide gas from volcanoes was broken up by ultraviolet light. The falling sulphur particles carried a sulphur isotope composition into rocks that indicates atmospheric formation. After the Great Oxidation, oxygen combined with atmospheric sulphur, preventing elemental particles from forming. The sky cleared and the isotopic signature vanishes from sedimentary rocks.

With the Great Oxidation, Earth's stratospheric ozone layer formed. This region of concentrated ozone between 20 and 30 kilometres in height shields the Earth's surface from harmful ultraviolet rays. If there were no ozone layer, you would be severely sunburned in tens of seconds. Ozone is a molecule of three oxygen atoms (O_3), which comes from oxygen. An oxygen molecule (O_2) is split by sunlight into oxygen atoms (O), which then combine with other O_2 molecules to make ozone (O_3).

The cause of the Great Oxidation is debated but must boil down to either a global increase in oxygen production from more organic burial or a decrease in oxygen consumption. The problem with the first idea is that organic matter extracts the light carbon isotope, carbon-12, from seawater, but seawater carbon, recorded in ancient limestones, doesn't steadily decrease in carbon-12 content. That leaves the second idea.

But what would have caused the flow of reductants to diminish? One possibility is that the relatively large abundances of hydrogen-bearing gases such as methane and hydrogen (which are inevitable in an atmosphere without oxygen) meant that hydrogen escaped rapidly from the upper atmosphere into space. The loss of hydrogen oxidized the solid planet. Oxidized rocks have fewer reductants so this gradually throttled further release of reductants. Oxidation happens because you can generally trace escaping hydrogen back to its source in water (H_2O), even if hydrogen atoms go through methane (CH_4) or hydrogen (H_2)

intermediaries. So when hydrogen leaves, oxygen gets left behind where the hydrogen originated. Water vapour itself has trouble reaching the upper atmosphere because it condenses into clouds, but other hydrogen-bearing gases such as methane don't condense and sneak the hydrogen out into space.

A rough analogy for the Great Oxidation is an acid-base titration of the sort performed in high school by dripping acid into an alkaline solution. Suddenly, the solution changes colour, typically from clear to red. The ancient atmosphere reached a similar transition from hydrogen rich to oxygen rich. Rather than acid base, we call such a titration a '*redox* titration' because it's a competition between *re*ductants such as hydrogen and *ox*idants such as oxygen.

A boring billion years ended by the advent of animal life

The ocean and land adjusted to the change of the Great Oxidation, and by about 1.8 Ga, atmospheric oxygen had settled down and remained between 0.2 and 2 per cent for an amazingly long time. The evolution towards complex life was slow perhaps because anoxic waters that were often rich in dissolved iron or hydrogen sulphide underlay a moderately oxygenated surface ocean. Such anoxic conditions are toxic for complex life. In fact, evolution was so sluggish that the interval from 1.8 to 0.8 Ga is called the *boring billion*. According to one scientific paper, 'never in the course of Earth's history did so little happen to so much for so long'!

Actually, several notable events did occur in the otherwise boring billion and just before it. Around 1.9 Ga, the first fossils of single organisms appear that are visible to the naked eye. These impressions of spiral coils of a few centimetres diameter (named *Grypania*) were probably seaweeds. In China, *acritarchs*, a type of fossil larger than 0.05 mm in size with organic-walls, occur at 1.8 Ga. Some are thought to have been algal cysts, which form when algae turn into inactive balls during dry periods. New lineages also emerged, including red algae at 1.2 Ga, which were likely capable of sexual reproduction. Today, we boil red algae descendants to make agar, which is used to thicken ice cream, and we use one type, nori, to wrap sushi. Because sex was discovered, the boring billion was not *that* boring.

Starting around 750 Ma, and culminating after fits and starts about 580 Ma, atmospheric oxygen rose a second time to levels exceeding 3 per cent, and the deep sea became fully oxygenated. Animal biomarkers and tiny fossils (possibly sponges) date from around 630 Ma. But it's only after 580 Ma that we find complex fossils from centimetres to metres in size. These are strange soft-bodied organisms without mouths or muscles that must have relied upon diffusion of nutrients through their skin. Some resemble pizzas while others look like plant-like fronds. However, the organisms lived too deeply in the sea to receive sunlight so they cannot have been plants. Collectively, they are called the *Ediacaran biota* because they were first clearly identified in the Ediacaran Hills in southern Australia. The Ediacarans lasted tens of millions of years before dying out.

Then, the *Cambrian Explosion* occurred, which was the relatively rapid appearance of animal fossils with new body architectures in a 10- to 30-million-year interval after the beginning of the Cambrian Period at 541 Ma. Representatives of most modern groups of animals emerged, including many with hard skeletons. The Cambrian (541–488 Ma) comes directly after the end of Proterozoic Aeon, and is the first period of the Phanerozoic Aeon (541 Ma to present). Phanerozoic means 'visible animals' from the Greek *phaneros* (visible) and *zoion* (animal). One tiny Cambrian creature, *Pikaia*, was a sort of swimming worm, a few centimetres long. *Pikaia* was probably in a group from which all the vertebrates, including us, evolved.

The second rise of oxygen allowed a diversity of animals to arise but its cause remains unsolved. Suggestions include organic carbon burial (and hence oxygen production) accelerated by several factors. These were new life on land that boosted the breakdown of surfaces and nutrient release, the evolution of marine zooplankton, and the enhanced production of clays that helped to bury organic matter. Alternatively, moderate levels of around 100 parts per million of atmospheric methane throughout the 'boring billion' might have promoted yet more hydrogen escape to space that oxidized the seafloor and ushered in an increase of oxygen after a second global redox titration.

Around 400 Ma, a third rise of oxygen took place when vascular plants colonized the continents. Such plants have special tissues (such as wood) to transport water and minerals. They probably enhanced nutrient release (by breaking down land surfaces with roots) and organic burial. Afterwards, atmospheric oxygen levels remained within 10 per cent to 30 per cent, allowing the persistence of animals and eventually our own appearance.

Snowball Earth or Waterbelt Earth?

Curiously, worldwide glaciations are associated with both the Great Oxidation Event and the second rise of oxygen, like bookends encompassing the 'boring billion'. Some scientists argue that at these times, the Earth's oceans entirely iced over in a so-called *Snowball Earth*. This state was far more extreme than the ice ages that have occurred in the last few million years that have only extended down to mid-latitudes, not the tropics. For context, the Proterozoic aeon is divided into three eras, the Palaeoproterozoic (2.5–1.6 Ga), Mesoproterozoic (1.6–1.0 Ga), and Neoproterozoic (1.0–0.541 Ga). Geologists commonly speak of the 'Palaeoproterozoic glaciations' to describe three or four glaciations that occurred at the beginning (2.45–2.22 Ga) of the Palaeoproterozoic, while the 'Neoproterozoic glaciations' refers to two large glaciations (at 715 and 635 Ma) and one small one (at 582 Ma).

The rock record provides evidence of these worldwide glaciations. The flow of ice sheets mixes stones, sand, and mud into a type of rock called *tillite*. Glaciers also leave parallel scratch marks called striations on underlying rocks. Furthermore, rocks drop out of icebergs or melting ice shelves and end up below as so-called *dropstones* that deform sedimentary beds. Together, tillites, *glacial striations*, and dropstones are dead giveaways that ice sheets were once present.

Magnetic measurements made in the 1980s and 1990s also demonstrated that the glaciations extended into the tropics. A compass shows the Earth's north–south magnetic field. Less obvious is the fact that the magnetic field lines are approximately perpendicular to the surface near the poles and parallel to the surface near the equator. Volcanic rocks contain iron minerals, such as magnetite (an iron oxide, Fe_3O_4), which become magnetized in the direction of the prevailing magnetic field when they cool. Thus, when the trapped or remnant magnetic field lines are parallel to ancient beds of rock, we know that the rocks formed in the tropics. This was discovered in rocks associated with Neoproterozoic and Paleoproterozoic glaciations.

The idea that the entire Earth might freeze over if sea ice reached the tropics originated in the 1960s. A Russian climatologist, Mikhail Budyko, calculated that if polar ice caps extended farther equatorward than 30° latitude, the entire Earth should freeze. Ice reflects a lot of sunlight and so has a high *albedo*, which is the fraction (between 0 and 1) reflected. Today, the Earth's albedo is about 0.3, meaning that 30 per cent of the sunlight is reflected to space. But sea ice reflects 50 per cent when bare or 70 per cent if covered in snow. A disastrous *ice-albedo runaway* occurs when high-albedo ice in the tropics makes the Earth absorb less sunlight and cool, which creates more ice that makes the Earth even colder, and so on. A frozen globe is created with an average temperature below −35°C and an ice thickness of 1.5 km at the equator and 3 km at the poles.

Joe Kirschvink of the California Institute of Technology proposed how a Snowball Earth would melt, as well as the very phrase *Snowball Earth*. He suggested that volcanoes would punch through the ice and vent carbon dioxide into the atmosphere. Because there would be no rainfall or significant open ocean to dissolve carbon dioxide, it would build up over a few million years until an enhanced greenhouse effect melted the Snowball. Also, the flow of ice from thick polar ice to thin tropical ice would carry volcanic ash and windblown dust, so that the tropics would become darker, making the Snowball prone to melt. Then the ice-albedo runaway would run in reverse: less ice would make the Earth warmer, which would melt more ice, and so on.

Layers of carbonate rock called *cap carbonates* that sit on top of the glacial deposits possibly attest to this aftermath of a Snowball Earth. The idea is that after the ice melted, large amounts of carbon dioxide were converted into cap carbonates. Proponents of the Snowball Earth hypothesis note that the ratio of carbon-12 to carbon-13 isotopes in the carbonate is similar to that in volcanic carbon dioxide.

A central problem for the Snowball Earth hypothesis is how photosynthetic algae survived under the thick ice, given that they require sunlight. Lineages of green and red algae went straight through the Neoproterozoic snowballs unscathed. One suggestion is that heat around volcanoes allowed some open water or thin, transparent ice. But others question whether the entire Earth really froze over. Some reasonably sophisticated computer simulations of the Earth's ancient climate produce a tropical belt of open water—a *Waterbelt Earth*—because this region is surrounded by bare ice of moderate albedo, whereas the very reflective, snow-covered ice occurs only at high latitudes.

Whatever happened, the cause of both the Neoproterozoic and Paleoproterozoic glaciations was probably a diminished greenhouse effect. Continents drift because of plate tectonics, and reconstructions have them bunched up in the tropics prior to the ancient snowball eras. Rainfall over tropical continents perhaps drew carbon dioxide down to lower levels than typical, poising the Earth for glaciation. Also, the coincidence of rises in oxygen with the glaciation eras suggests a trigger through methane, another greenhouse gas. Probably the oxygen increases were associated with rapid decreases in methane because high abundances of the two gases are incompatible—they react.

For astrobiology, the lesson from Snowball Earth is that the biosphere on an Earth-like planet is resilient. Despite drastic climate swings, life on Earth survived and emerged after the Neoproterozoic into a new aeon of animals.

On the occurrence of advanced life

In astrobiology, we often wonder if animal-like life exists elsewhere in the galaxy. That possibility can perhaps be informed by evolution on Earth. To become complex, terrestrial life had to overcome at least two key hurdles. First, life had to acquire complex cells, which were needed for differentiation on the larger scale, i.e. cells with functional distinctions, such as liver versus brain cells. Large, three-dimensional multicellular animals and plants are made only of *eukaryotic* cells, which can develop into a much larger range of cell types compared to the cells of microbes (see Chapter 5).

A second precursor for animals was having sufficient oxygen to generate lots of energy with an *aerobic* (oxygen-using) metabolism, such as our own. Abundant energy is needed to grow large and move. *Anaerobic* metabolism, which doesn't use oxygen, produces about ten times less energy for the same food intake than aerobic metabolism. For extraterrestrial life, we might wonder if fluorine or chlorine—which are powerful oxidants—could be used in place of oxygen to generate high energy levels. The answer is no. Fluorine is so reactive that it explodes with organic matter, while chlorine dissolves to make harmful bleach. Oxygen, it appears, is the best elixir for complex life in the periodic table—reactive enough but not too aggressive.

Oxygen not only needs to be present but also concentrated. The first aerobic multicellular life probably consisted of aggregates of cells produced when dividing cells failed to separate. Such agglomerations would have been limited in size by diffusion of oxygen to the innermost cell so that greater levels of external oxygen would have permitted larger cell groupings. At typical metabolic rates, a diffusion-constrained animal precursor of about a few millimetres' dimension would need atmospheric oxygen concentrations exceeding 3 per cent. This corresponds to the oxygen increase around 580 Ma that ushered in the animals.

Fossil algae show that eukaryotes existed long before animals, so the slow fuse to the Cambrian Explosion was probably the time it took to build up adequate oxygen. We can even define an *oxygenation time* as that required to reach oxygen levels sufficient for complex animal life. On Earth, it was 4 billion years after the planet formed.

The oxygenation time on Earth-like exoplanets is uncertain given that we're still trying to understand what set the timescale for the Earth. However, oxygen comes from liquid water. So Earth-like exoplanets with oceans have the potential to develop oxygen-rich atmospheres if water-splitting photosynthesis evolves. If the oxygenation time exceeds 12 billion years on a certain exoplanet because it's endowed with material that reacts with oxygen, a Sun-like parent star would turn into a 'red giant' before the conditions for animal life became possible, and the planet would be doomed to possess nothing more complex than microbial slime. But if other exoplanets have short oxygenation times, complex life might occur faster than it took on Earth.

Trends in evolution? The lesson of mass extinctions

Once animals evolve, we might wonder if evolution has a trend towards greater complexity. Consider life on land. In land colonization, the first plant spores are found around 470 Ma, while fossils of substantial parts of plants appear about 425 Ma. Insects appear on land around 400 Ma. Then fish evolved into amphibians around 365 Ma and their descendants ultimately became the reptiles and mammals, including us. This is a trend for the biosphere to use more of the Earth's resources. But a misconception is to think of evolution as a steady progression ending in us. The fossil record shows that species come and go. Also, mass extinctions intermittently prune the diversity of complex life. They are events when more than 25 per cent of families are lost, where 'family' is the biological tier above genus and species (see Chapter 5). Such mass extinctions refer to the non-microbial part of the biosphere, of course.

Only around the Proterozoic–Phanerozoic transition is there a dramatic increase in diversity because at that time organisms acquired new body architectures and ways of living that persisted. In animals, a key innovation was the body cavity, the *coelom* (pronounced 'see-lum'), which could be filled with fluid to provide rigidity and allow the concentration of forces from muscles. This facilitated self-propulsion, which was further improved with hard skeletons, at first external and then internal. A predator–prey evolutionary arms race likely contributed to the Cambrian Explosion of animal varieties. Plant diversity later increased with the development of vascular systems with rigid cells for transport and associated organs that eventually gave rise to trees.

In the past 500 Ma, there have been five mass extinctions that killed over 50 per cent of species, with the two biggest at 251 Ma and 65 Ma, respectively. Excluding crocodilians, no land animal larger than the size of a domestic dog got through the largest event, separating the Permian (299–251 Ma) and Triassic

(251–200 Ma) periods. This event took a massive toll in the oceans. For example, the trilobites—icons of the Cambrian seas—had been in decline but the Permian–Triassic extinction finished them off. The second largest event at 65 Ma wiped out the dinosaurs and separates the Cretaceous (145–65 Ma) and Paleogene (65–23 Ma) periods. (Some literature refers to this event as Cretaceous–Tertiary based on an older naming system.) The Permian–Triassic mass extinction destroyed as much as 95 per cent of marine species and 70–80 per cent of land animals, while the Cretaceous–Paleogene extinguished 65–75 per cent of all species.

The cause of the Permian–Triassic mass extinction was apparently a chain of misfortunes generated by the Earth herself. The trigger seems to have been large-scale volcanism in Siberia, covering an area similar to Europe. Because coal deposits underlay this area, the volcanism pumped out huge amounts of coal-derived methane and carbon dioxide greenhouse gases, which warmed and acidified the ocean. Oxygen is less soluble in warm water, so an anoxic deep sea developed, which may have belched poisonous hydrogen sulphide to the surface. The combination of climatic warming, ocean acidification, and noxious gases extinguished more life than any other event in the Phanerozoic.

The impact of an asteroid of about 10 kilometres diameter appears to have caused the Creataceous–Paleogene mass extinction. Calamitous consequences included temporary destruction of the ozone layer, worldwide wildfires, acid rain, and subsequent climatic cooling caused by sulphate particles injected into the stratosphere that reflected sunlight. The impact crater has even been found about a kilometre underground near the Mexican town of Chicxulub (pronounced 'sheek-soo-loob').

Extinctions destroy previously successful lineages but they also provide opportunities for others. You are reading this book because of the Chicxulub impactor. The mammals became dominant once the dinosaurs were gone.

On the other hand, mass extinction events mean that if a civilization develops, it can be wiped out. Impacts the size of that producing the Cretaceous–Paleogene extinction should occur every 100 million years or so on Earth, give or take a factor of a few. The implication is that civilizations on exoplanets could be short-lived compared with the age of the universe merely because of such random catastrophes. Thus, an Earth-like planet can remain fit for life, but civilizations are probably ephemeral. The latter has consequences in the astronomical search for extraterrestrial intelligence (Chapter 7).

Chapter 5
Life: a genome's way of making more and fitter genomes

Life on Earth: the view from above

In Chapter 1, I presented a general definition of life, but in finding life elsewhere, it helps to know our one example of life in great detail. To this end, let's start with a global perspective of terrestrial biology and work downward to cells and molecules.

Imagine an interstellar traveller who arrives on the Earth and wants to know about our biology. Perhaps from her planet, she had deduced that life exists here because of Earth's wet, anomalously oxygen-rich atmosphere, or she had picked up TV broadcasts from decades ago that somehow didn't put her off visiting. Amazingly, she speaks English (as extraterrestrials always do in the movies) and by a stroke of luck her spacecraft lands in an English-speaking country! What would we tell her?

At the global level, Earth's *biosphere* is the sum of all living and dead organisms. Sometimes the term includes the non-living regions that life occupies. By quantifying the biosphere in billions of tonnes of carbon (1 tonne = 1,000 kg), we can identify its broad components. The biomass on land is around 2,000 billion tonnes of carbon of which 30–50 per cent is living and the rest is dead. In the ocean, only 0.1–0.2 per cent of about 1,000 billion tonnes of biomass carbon is alive. Forests are the reason that there's so much more living biomass on land than in the oceans.

An unsettled issue that's relevant to life elsewhere (below the surface of Mars or Jupiter's moon, Europa) is the extent of Earth's *subsurface biosphere* or 'intraterrestrial life.' Some scientists suggest that a huge mass of microbes extends a kilometre or two below the seafloor and more than 3 kilometres underneath land. A limit for life at such depths is temperature. As you go downwards, it gets warmer (as miners know) and, at some point, too hot for even the toughest microbes. Earth's subsurface biosphere biomass is uncertain because deep drilling has not been done for all types of subsurface environments, but estimates range from about 1 per cent to 30 per cent of Earth's living biomass.

Whatever its precise mass, the biosphere is less than a billionth of the mass of the Earth and yet manages to greatly influence the chemistry of the surface environment. The biosphere can do this because it processes vast amounts of material with rapid turnover. In doing so, individuals live and die almost instantly on geological timescales. Microbes typically reproduce in tens of minutes to days, while large multicellular organisms last only a few thousand years at most before they become dead fodder for microbial degradation. For example, amongst the latter, the oldest non-clonal organism is a Bristlecone Pine in the White Mountains of California, which germinated around 3049 BC, several centuries before the earliest Egyptian pyramids, but a mere blip in geological time.

The main activities in today's biosphere are oxygenic photosynthesis and its chemical reversal through aerobic respiration and oxidation (Chapter 4). However, microbes possess a huge range of other metabolisms, which we discuss later.

Microbes, of course, are found almost everywhere at the Earth's surface. A seawater and freshwater bacterium called *Pelagibacter ubique* is probably the most numerous organism on Earth. Despite that, it was first described only in 2002, which shows how biology is still developing. Overall, there are about 10^{29} microbes of all varieties in the ocean, far exceeding the 10^{22} stars in the observable universe. Microbes are also abundant on land. There are typically about 100 million to 10 billion microbes per gramme of topsoil. Indoor air usually has about a million bacteria per cubic metre. An average of one microbe (albeit dead) even floats in every 55 cubic metres of air at 32 kilometres altitude in the stratosphere.

There are four key properties that have allowed the Earth to become so extensively inhabited. The most important is widespread liquid water. All metabolizing organisms contain organic molecules dispersed in aqueous solution. Consequently, we don't expect life to exist on a completely dry surface, such as Venus. The second essential property is having energy for metabolism. Sunlight is the main source for Earth's biosphere, but some organisms obtain energy from chemical reactions in darkness, which means that it's not impossible that microbial-like life might exist below the surface of Mars or Europa. A third life-giving feature is a renewable supply of essential chemical elements. A planet that can't resupply vital elements through natural cycles (such as the water cycle or tectonics) would be deathly. A fourth attribute, which may be essential for life, is the presence of interfaces between solids, liquids, and gases. It's advantageous to live at a stable interface, such as on land or the surface of the ocean. This is why it would be difficult for life to exist on a gas giant like Jupiter that has no surface. Life might speculatively be possible at certain altitudes in Jupiter's atmosphere but deep churning by convection would periodically plunge life into an interior of fatal heat and pressure.

An inside view of terrestrial life: the cell

We have just considered biology on a global scale but under the microscope all organisms are composed of cells of different types that put them into one of three domains, the *Eukarya, Archaea*, or *Bacteria*. The last two are microbial and are sometimes lumped together as *prokaryotes*, although many microbiologists now consider this term antiquated because archaea and bacteria are biochemically dissimilar. DNA floats freely in the middle of the archaeal and bacterial cells, whereas in eukaryotes the DNA is housed inside a membrane-bound nucleus. Archaea and bacteria are single cells, with the exception of some bacterial species that join up in a row to form filaments. Eukaryotes can be single celled, such as an amoeba or a baker's yeast, but only eukaryotes form large, three-dimensional multicellular organisms, such as mushrooms or humans.

The classification into three domains of life was motivated by genetics and supersedes an older 'five kingdom' system of plants, animals, fungi, protists (single-celled eukaryotes), and bacteria. However, these old terms are still used in taxonomy, which classifies an organism below its domain according to Kingdom, Phylum, Class, Order, Family, Genus, and Species. The mnemonic I use to remember these levels (or taxa) is far too rude to mention but another is 'Keeping Precious Creatures Organized For Grumpy Scientists'. The taxonomic levels of a human, for example, are the animal kingdom, chordate phylum, mammal class, primate order, hominid family, *Homo* genus, and *sapiens* species. The Swedish botanist Carolus Linnaeus (1707–78), who gave us the word *biology*, developed modern taxonomy, including binomial names for organisms, e.g. *Homo sapiens*. But it took the advent of evolutionary theory and molecular biology to uncover the biochemical unity of life and genetic common ancestry.

While there are some similarities between eukaryotes and the other two domains, there's also a gulf in complexity. Bacteria and archaea are usually around 0.2–5 microns (millionths of a metre) in size, with rare exceptions, whereas eukaryotic cells are generally bigger, at 10–100 microns size. The larger eukaryotic cells contain organelles to perform specialized functions, analogous to the

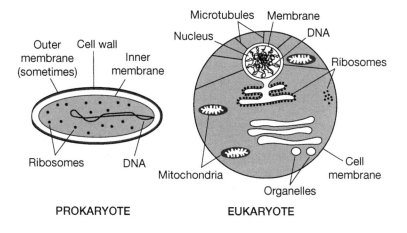

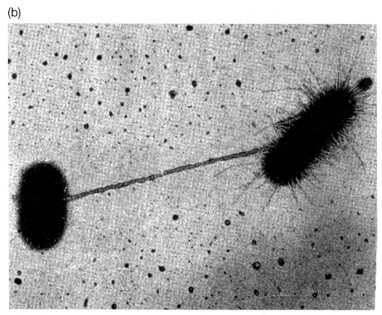

4. a) Schematic of prokaryote (archaea and bacteria) versus eukaryote structure; **b)** Two bacteria caught in the act of conjugation

organs of the human body (Fig. 4). For example, the *mitochondria* carry out respiration. In plant or algal cells, *chloroplasts* perform photosynthesis. However, one feature common to all cells is a large number of ribosomes, which are globular structures that make proteins. For example, in prokaryotes or simple eukaryotes such as yeast, the cell might have several thousand ribosomes, while in an animal cell the number might reach several million.

Given an astrobiological interest in complex extraterrestrial life, we might ask why only the eukaryotic cell produces large, three-dimensional multicellular life. The answer is not fully known but eukaryotic cells have a more sophisticated internal dynamic cell skeleton or cytoskeleton than archaea or bacteria. This consists of protein microfilaments, tiny protein tubes ('microtubules'), and molecular motors that control cell structure and help transport signalling molecules to change cell physiology. So the ability to develop into many specialized forms (for example, skin or brain cells) is inherent in the make-up of a eukaryotic cell. Cyanobacteria are able to make filaments of hundreds of cells in a row with some different cell types but that is the limit. Unlike archaea or bacteria, eukaryotes have bigger and more modular genomes, which also allows for more complexity. If there were no eukaryotic cells, the Earth would be much duller. All of the familiar organisms of our world—the animals, plants, and fungi—wouldn't exist. Thus, when we think about complex life on exoplanets, we should wonder whether evolution would make cells like eukaryotes elsewhere. For this reason, the origin of eukaryotes is of great interest.

Modern genetics implies that the eukaryotic cell is a 'Frankenstein's monster', assembled in evolutionary history from bits and pieces of bacteria and archaea. For example, the mitochondrion in eukaryotic cells was derived from a bacterium originally living symbiotically inside another cell. The larger cell, which may have been an archaeon or some 'proto-eukaryote' that no longer exists, swallowed the free-living bacterium and, in the most important gulp in history, the mitochondrial ancestor came into being. Indeed, separate DNA in mitochondria provides evidence of bacterial ancestry. The distinct DNA of chloroplasts in plant and algal cells shows that chloroplasts were derived in a similar way from symbiotic cyanobacteria that ended up living inside larger cells. Effectively, all the cells in the leaves of green plants contain ancestors of cyanobacterial slaves that were caged and co-opted long ago. The theory of such an origin for the mitochondria, chloroplasts, and other organelles in eukaryotic cells is called *endosymbiosis*.

A world without eukaryotes would also be one without sex. Think no flowers or love songs. Archaea and bacteria are not sexed but they do conjugate when two cells are connected by a tube in which genes are transferred in pieces of DNA called plasmids. Bacterial surfaces have protuberances called pili, and during conjugation a special pilus extends to a partner, providing the conduit (Fig. 4). Unlike sexual reproduction in eukaryotes, microbial conjugation doesn't produce offspring and is quick and easy. It's as if you brushed up against a person with red hair in a coffee shop and acquired the red-haired gene with an instant change of your hair colour. Archaea and bacteria can also acquire new genes through transformation (uptake of foreign DNA from the environment) and transduction (virus-mediated gene swapping). Indeed, the rapid acquisition of genes allows the quick development of bacterial resistance to antibiotics.

Eukaryotes that are sexed generate *gametes,* i.e. sperm and egg cells. In complex multicellular eukaryotes, the diversity of life makes it surprisingly hard to define 'male' and 'female', leaving us with the odd definition that males are those that produce the small gametes, while females produce the big ones. The gametes fuse so that half of the genes come from a father and half from a mother. Usually DNA is a loosely coiled thread but before cell division, DNA curls up into visible *chromosomes* under a microscope. For example, humans have 46 chromosomes in 23 pairs in each cell, except for the *gamete*s, which have half, i.e. 23 chromosomes.

There are many ideas about why sex is evolutionarily advantageous for eukaryotes. One possibility concerns how it mixes and matches genes from both parents onto each chromosome in a process called *recombination*. If beneficial mutations occur separately in two individuals, the mixture of both can't be achieved in asexual organisms, but sexually reproducing organisms can bring them together and reap the benefits. Conversely, sex can also eliminate bad, mutated genes by bringing unmutated genes together in some individuals, whereas self-cloning organisms are stuck with bad genes, and offspring can die because of them.

Outside the three domains, viruses represent a grey area between the living and non-living. Viruses are typically about ten times more abundant than microbes in seawater or soil. They consist of pieces of DNA or RNA surrounded by protein and, in some cases, a further membrane. Viruses are tiny, only about 50–450 nanometres (billionths of a metre) in size, comparable to the wavelength of ultraviolet light. They are generally considered non-living because they are inanimate outside a cell and have to infect and hijack cells for their own reproduction. However, some do this without the host ever noticing, so not all viruses cause disease. One theory of several for the origin of the nucleus of eukaryotes is that it may have evolved from a large DNA virus, but the role of viruses in the evolution of life is still a matter of debate.

The chemistry of life

To discuss many aspects of life, such as genetics and metabolism, requires the vocabulary of biochemistry. The four main classes of biomolecule are nucleic acids, carbohydrates, proteins, and lipids.

Like self-assembly furniture, many biomolecules are modular. They are chains or polymers of smaller units called monomers.

Carbohydrates provide energy and structure. They contain C, H, and O atoms in a 1:2:1 ratio, with repeated units of $C(H_2O)$, so literally, their chemical composition is 'carbon hydrated'. Sugars with five carbon atoms are found in DNA and RNA molecules, while six-carbon sugars exist in cell walls, such as cellulose in plants.

Lipids are organic molecules that are insoluble in water but dissolve in a non-polar organic solvent—one without significant electrical charge on any of its atoms, such as olive oil. Thus, if we ground up a dead animal and bathed it in a non-polar solvent, anything that dissolved would be a lipid. Life uses lipids in cell membranes, as fats for energy storage, and as signalling molecules. Major components of membranes are phospholipids, which have a hydrophilic (water-loving) end that contains phosphorus and a hydrophobic (water-repellent) tail that is a hydrocarbon, consisting of carbon and hydrogen atoms. A double layer of phospholipids, called a *bilayer*, forms a membrane. The hydrophilic ends stick out into an aqueous medium on the interior and exterior of a cell, while hydrophobic tails face each other in the middle of the membrane.

Proteins are polymers made up of amino acid units. Their use includes enzymes and structural molecules, although there is a long list of other functions.

RNA and DNA are nucleic acids, which are polymers of *nucleotide* monomers. Each nucleotide is made up of a five-carbon sugar, a phosphate, and a part called a base (Fig. 5). In DNA, there are four possible bases. All of them contain one or two rings of six atoms where four of the atoms are carbon and two are nitrogen. Each base has a letter designation of A, C, G, and T, which stand for adenine, cytosine, guanine, and thymine molecules,

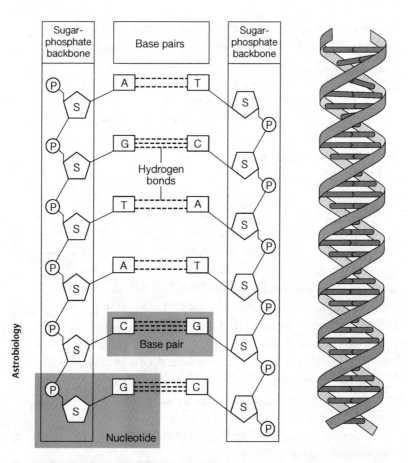

5. *Left*: DNA consists of two strands connected together. Each strand is made up of a 'backbone' of phosphate (P) and sugar (S) components. The strands connect to each other with base pairs. *Right*: In three dimensions, each strand is a helix, so that overall we have a 'double helix'

respectively. Molecules of RNA use the same three bases except that the T of DNA is replaced by a U for uracil.

In 1953, James Watson and Francis Crick famously deduced the structure of DNA: two polynucleotide strands coiled in a screw-like helix (Fig. 5). The bases at the sides of each strand stick together by 'hydrogen bonds' in which a slightly positively charged hydrogen atom on one base is attracted to a slightly negatively charged atom on a base from the opposite strand. Structural compatibility only allows a C base to pair with G and an A to pair with T. Each DNA molecule has several million nucleotides, and each chromosome in the cell contains a DNA double helix. In contrast, RNA molecules are mostly single stranded. However, RNA can fold back on itself if complementary bases exist in two separate parts of the strand, noting that adenine pairs with uracil (U) in RNA.

The structure of DNA incorporates two fundamental characteristics of life identified in Chapter 1: an ability to reproduce; and a blueprint for development and maintenance. In replication, the DNA helix splits into two strands with each serving as a template for a new complementary strand. For example, wherever A appears on the template, a T is added to the newly generated strand, or vice versa. The same applies for G–C pairs. In the process, mutations and mistakes allow for evolution.

In fulfilling its other role as the blueprint, the DNA double helix is unzipped by an enzyme to provide instructions to generate proteins. Part of the unzipped DNA undergoes *transcription* into a strand of messenger RNA (mRNA), which is a complementary copy of the DNA, except that U (instead of T) is inserted wherever A appears in the DNA. Then the mRNA is fed into a ribosome like a ribbon. In the *genetic code*, groups of three letters (called *codons*) along the mRNA specify each amino acid in a protein that flows out of the ribosome.

Apart from reproduction, life has to sustain itself through metabolism, which involves breaking down molecules to make energy (*catabolism*) as well as building up biomolecules (*anabolism*). The classification of metabolisms depends on the need for energy and carbon. Because each of these requirements can, in turn, be satisfied in two different ways, biologists have $2 \times 2 = 4$ metabolic terms for organisms: *chemoheterotroph, chemoautotroph, photoheterotroph,* and *photoautotroph* (Fig. 6).

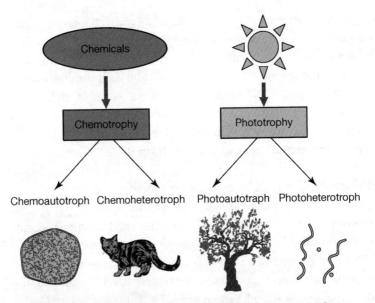

6. The classification scheme for metabolisms in terrestrial life

All organisms (probably even extraterrestrial ones) fall into one or more of these categories. The *troph* suffix means 'to feed' and the two ways in which organisms get energy, from chemical sources or sunlight, give rise to the *chemo-* and *photo-* prefixes. An additional *hetero-* or *auto-* prefix is employed depending upon the method used to acquire carbon. If carbon is obtained by consuming organic carbon compounds (such as sugars), the *hetero-* prefix applies. If an organism converts inorganic carbon (e.g. carbon dioxide) into organic carbon—called 'fixing carbon'—the *auto-* prefix is used. In general, heterotrophs must acquire food to make energy, while autotrophs can fix carbon and make their own energy.

We humans are chemoheterotrophs. You and I consume organic chemicals made by other organisms, such as plants. All animals and fungi, many protists, and most known microbes are chemoheterotrophs. In contrast, a chemoautotroph is a microbe that uses inorganic chemicals such as hydrogen, hydrogen sulphide, iron, or ammonia to make energy, some of which it uses to extract carbon from carbon dioxide. For example, chemoautotrophic microbes live in darkness in deep-sea hydrothermal vents by oxidizing hydrogen sulphide and other substances. (Chemoautotrophs are also called *chemolithotrophs*, from the Greek 'lithos' for stone because the inorganic chemicals they use come from geological sources.) Plants, algae, and some cyanobacteria are all photoautotrophs because they use sunlight for energy and acquire carbon from the air. Photosynthetic bacteria from the *Chloroflexus* genus, which are found in hot springs, are an example of microbes that metabolize as photoheterotrophs using sunlight for energy and acquiring carbon from organic compounds made by other microbes. Chemoautotrophs might inhabit the subsurface of Mars or Europa, while phototrophs might exist in the oceans of habitable exoplanets.

The tree (or web) of life

The history of life on Earth can be deduced from the way that evolution has altered genes. Evolution is the change in inherited characteristics in a population from one generation to the next. Because individuals are genetically variable, in any given environment some will be better adapted and have greater reproductive success than others, which biologists describe as higher *fitness*. In every generation, individuals of lower fitness are lost. This is natural selection. So, over many generations, lineages accumulate genetic adaptations and new species evolve.

In sexual organisms, *species* are groups that cannot interbreed under natural conditions, such as humans and horses. Many bacteria and archaea live in different ecological niches that define separate groups, but the ease with which genes can move between microbes makes it more difficult to designate species of microbe. Consequently, genetics is used. A difference of about 2–3 per cent in the genes that code the RNA contained in ribosomes is enough to separate microbial species.

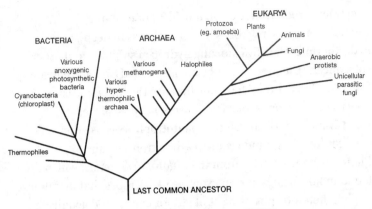

7. The 'tree of life' constructed from ribosomal RNA

In fact, we can assess the relatedness of all life forms by comparing genes. For example, genes indicate that you and the fungus between your toes—which are both eukaryotes—are much closer relatives than an archaeon and a bacterium, despite appearances. From the mid 1960s onwards, various techniques have been developed to assess organisms at the molecular level, including a comparison of the sequences of either amino acids in proteins or nucleotides in RNA and DNA.

Genes define the sequence of amino acids in a protein, so differences in the 'protein sequence' between species indicate disparity or relatedness. For example, 'cytochrome c' is a respiratory protein of 104 amino acids found in many organisms. The sequences are identical in humans and chimps, showing their close relationship. Compared to the human protein, a rhesus monkey has one different amino acid, a dog has thirteen different amino acids, and so on. As species become more distantly related, the protein sequences diverge. With this sort of data, an analyst can draw up a tree, like a family tree, that relates all the species.

In the 1970s, the American microbiologist Carl Woese (1928–2012) used nucleic acids rather than proteins to determine relationships among organisms. He examined RNA in ribosomes (rRNA, for short), with the key insight that this enabled analyses of all life. Protein synthesis, the job of ribosomes, is a core function of any cell. Consequently, the genes that code for rRNA should have mutated slowly over time because most mutations for this process would have been fatal and so failed to accumulate. About 60 per cent of the dry weight of ribosomes is made up of rRNA, and the remainder is protein. Ribosomes in prokaryotes are smaller than in eukaryotes, but otherwise similar in structure and function, allowing their comparison.

Woese isolated a small rRNA and by comparing the differences in the nucleotide sequences, he built a 'tree of life' (Fig. 7). The construction of the evolutionary history of organisms in this way is called *phylogeny*. Woese's tree was a shock because life grouped into the three domains mentioned earlier, unlike the 'five kingdom' paradigm in which archaea were lumped with the bacteria. Furthermore, Woese's tree demoted plants, animals, and fungi to mere twigs at the end of the eukaryotic branch.

Today, while Woese's conception of three domains is widely accepted, studies of other genes show that a tree with vertical descent from one generation to the next is too simple. This is because microbes swap genes willy-nilly (Fig. 4), sometimes to unrelated species, which is called *lateral* (or *horizontal*) *gene transfer*. Thus, microbial branches of the tree of life are more like a 'web of life', criss-crossed by lateral gene transfers.

Since the mid 1990s, DNA sequencing has become more automated. For sequencing a single gene from environmental samples, the process is as follows: isolate DNA from cells → copy (or 'amplify') a DNA gene many times using a procedure called the 'polymerase chain reaction' → obtain the gene sequence → compare with other organisms → produce a phylogenetic tree. Increasingly, whole genomes are sequenced, including the human genome, which contains roughly 21,000 protein-coding genes.

Astonishingly, human protein-encoding genes cover only 1.5 per cent of 3 billion nucleotides. The rest of the sequence is said to be 'non-coding' or 'junk DNA'. These terms are actually misnomers because many parts of non-coding DNA regulate when certain genes are expressed (i.e. give rise to proteins) or code for non-protein products such as rRNAs.

The surprising division between archaea and bacteria revealed by genes is confirmed by biochemical differences. For example, bacterial cell walls contain peptidoglycan, which consists of carbohydrate rods cross-linked by proteins, whereas archaeal cell walls have variable chemistry of protein or carbohydrate, or both. Eukaryote cell walls, for comparison, are made of cellulose in plants, chitin in fungi, and are non-existent in animal cells, which have only a membrane. Furthermore, the cell membrane lipids of bacteria and archaea differ. First, there are dissimilar chemical linkages between the hydrophobic and hydrophilic ends of phospholipids. Second, instead of a lipid bilayer with hydrophobic ends dangling face-to-face in the middle of the membrane, archaea have molecules that connect all the way through. This provides strength and is one reason why some archaea can live in very hot water as hyperthermophiles. There are also ecological distinctions. For example, no archaeon is a pathogen, so you'll never get a disease from archaea whereas you can from various bacteria.

One further use of phylogeny is that the genetic difference between taxa can be related to the time in geological history that they diverged, making a *molecular clock*. In the 1960s, Emile Zuckerkandl and the Nobel Prize-winning chemist Linus Pauling compared proteins for taxa that were known from fossils to have diverged from a common ancestor. They found that the number of amino acid differences is proportional to elapsed time. One interpretation is that most changes are 'neutral mutations' that have no effect on fitness. Similarly, the number of nucleotide substitutions in certain DNA sequences is proportional to elapsed time. Molecular clocks work best with closely related groups of species that are likely to have had similar rates of mutation. Distantly related species have disparate generation times and metabolic rates with variable mutation rates that must be taken into account. To use a molecular clock, a calibration point from the fossil record fixes the date of a particular ancestor in a computer algorithm applied to the molecular data. As we go back very deep into time, there are fewer fossils, so the technique becomes challenging. Nonetheless, molecular clocks indicate that the last common ancestor of animals occurred about 750–800 Ma, which is curious because it predates the oldest animal fossils.

Despite the complications of lateral gene transfers, the tree of life provides information about early life. Thermophiles (microbes that thrive at high temperatures) that grow best at 80–110°C are found near the root of the rRNA tree in the archaea and bacteria (Fig. 7). A reasonable inference is that the last common ancestor lived in a hydrothermal environment. Organisms close to the root are also chemoautotrophs, suggesting that primitive microbes probably gained energy from inorganic compounds. The tree also shows that complex organisms—the plants, fungi, and animals—were late to evolve, consistent with the fossil record. Consequently, if Earth's phylogeny is a guide for astrobiology, the implication is the same as the fossil record: life elsewhere ought to be mostly microbial for much of the history of the host planet.

Life in extreme environments

Apart from being clustered near the last common ancestor, thermophiles were the starting point for research into *extremophiles*. Extremophiles are organisms that thrive under environmental conditions that are extreme from a human perspective. In 1965, Thomas Brock, an American microbiologist, discovered pink filaments of bacteria living at temperatures of 82–88°C in a steamy hot spring in Yellowstone National Park. At that time, no life was known to exist above 73°C, so Brock's discovery spurred an interest in exploring the limits of life.

Although it wasn't anticipated, Brock's efforts also ultimately enabled the explosion in genetics. Brock found a new bacterium, *Thermus aquaticus*, in another hot spring. From this microbe, industrial scientists isolated an enzyme stable at high temperatures that was able to catalyse the polymerase chain reaction (PCR)—the DNA duplication technique that revolutionized biology. Thus, pure science—in this case, what we now call astrobiology—ended up benefiting society unexpectedly. Today, PCR technology is a multibillion-dollar industry. Brock, however, gave away his bacterium and didn't get a penny.

Often, the growth (meaning replication) of extremophiles either requires extreme conditions or is optimal at them. For temperature, the upper limit for eukaryotes is 62°C, compared to 95°C for bacteria and 122°C for archaea. The record holder, the methane-producing archaeon, or methanogen, *Methanopyrus kandleri*, has optimal growth at 98°C.

Identification of various extremophiles (Box) indicates that life exists in a much wider range of environments than anyone thought feasible fifty years ago, which opens up possibilities for life beyond Earth. For example, thermophiles might survive deep underground on Mars because they exist deep in the Earth's crust. Another example concerns life in Lake Vostok (250 km long, 50 km wide, and 1.2 km deep), which sits below 4 kilometres of ice in east Antarctica. Ice some 100 metres above the lake is thought to have frozen from lake water. Oddly, it contains traces of the DNA of chemoautotrophic thermophiles. This suggests that beneath the cold lake (at −2°C because of the high pressure), there may be hydrothermal water containing

> **Box Extremophiles**
>
> *Acidophiles*: require an acidic medium at pH 3 or below for growth; some tolerate a pH of below 0.
>
> *Alkaliphiles*: require an alkaline medium above pH 9 for optimal growth and some live up to pH 12.
>
> *Barophiles (or piezophiles)*: live optimally at high pressure; some live at over 1,000 times the Earth's atmospheric pressure.
>
> *Endoliths*: live inside the pore space of a rock (*endo* = within, *lithos* = stone).
>
> *Halophiles*: require a salty medium, at least a third as salty as seawater (*halo* = salt).
>
> *Hypoliths*: live under stones (*hypo* = under).
>
> *Psychrophiles*: grow best below 15°C, while some grow down to –15°C, with reports as low as –35°C for metabolism.
>
> *Radioresistant microbes*: resist ionizing radiation, such as from radioactive materials.
>
> *Thermophiles*: thrive at temperatures between 60 and 80°C, while a *hyperthemophile* has optimal growth above 80°C; they can be contrasted with *mesophiles*, such as humans, which live between 15 and 50°C.
>
> *Xerophiles*: grow with very little available water, and may be halophiles or endoliths (*xeros* = dry).

thermophiles emanating from fractures. The techniques honed for detecting life in lakes such as Vostok can be applied in the search for life elsewhere in the Solar System. Sampling for life in the subsurface ocean of Jupiter's moon, Europa, involves similar challenges.

Chapter 6
Life in the Solar System

Which worlds might be habitable today?

In 2002, while teaching astrobiology, I offered a prize to any student who could guess nine celestial bodies up to the orbit of Pluto that I reckoned might possibly harbour extraterrestrial life (Table 1). No one won. But, with the growth of astrobiology discoveries and online information, someone received the prize in 2010. Today I might add several more bodies, but we'll leave those until the end of this Chapter.

The type of life that I'm considering in Table 1 is simple, comparable to microbes, and the guiding principle concerns liquid water. On Earth, wherever we find liquid water, we find life, whether in bubbling hot springs, drops of brine inside ice, or films of water around minerals deep in the crust.

Some go further and speculate that *weird life*—organisms that don't depend on water—might exist in lakes on Titan, the largest moon of Saturn. Titan's lakes contain liquid hydrocarbons, not water, somewhat like small seas of petroleum. Life that uses hydrocarbon solvent is unknown, and there are some arguments from physical chemistry, which I'll mention later, that such life might be difficult. In contrast, there's no question that liquid water can support life.

Table 1. Nine abodes where life might exist in the Solar System today. The distance to the Sun is in Astronomical Units (AU), where 1 AU is the Earth-Sun distance of about 150 million kilometres or 93 million miles

Body	Type of body and average distance from the Sun	Why it might have life
Mars	planet, 1.5 AU	might have subsurface pockets of liquid water
Ceres	largest asteroid, 2.8 AU	might have a subsurface ocean
Europa, Ganymede, Callisto	large icy moons of Jupiter, 5.2 AU	evidence for subsurface oceans
Enceladus	icy moon of Saturn, 9.6 AU	evidence for a subsurface ocean or sea and presence of organics
Titan	largest moon of Saturn, 9.6 AU	evidence for a subsurface ocean and presence of organics
Triton	largest moon of Neptune, 30.1 AU	might have a subsurface ocean
Pluto	large Kuiper Belt object, 39.3 AU	might have a subsurface ocean

Sunlight and the habitability of the inner planets

Whether the inner planets—Mercury, Venus, Earth, or Mars—have liquid water mostly has to do with temperature and, thus, distance from the Sun (Table 2). 'Location, location, location' doesn't just sell houses but is critical for planetary habitability.

Table 2. The inner planets and factors that affect their current habitability

	Mercury	Venus	Earth	Mars
Average distance to Sun (Astronomical Units)	0.4	0.72	1	1.5
Average surface temperature (°C)	167	462	15	−56
Greenhouse effect (°C)	0	507	33	7
Atmospheric pressure (bar)	0	93	1	0.006
Liquid water?	Too hot	Too hot	Abundant	Mostly too cold
Main gases in the atmosphere	airless	96.5% carbon dioxide, 3.5% nitrogen	78% nitrogen, 21% oxygen	95.3% carbon dioxide, 2.7% nitrogen

Location matters because sunlight spreads into a sphere with a surface area that grows with the square of the planet–Sun distance. *Solar flux* is the wattage (like a light bulb) of sunlight received per square metre. At the Earth's orbital distance of 1 Astronomical Unit (AU), sunshine provides 1,366 Watts over every square metre, the equivalent of almost fourteen 100-Watt light bulbs. At the 5 AU distance of Jupiter, the solar flux is a factor of 25 times smaller because the same energy spreads over a sphere that has $5 \times 5 = 25$ times the area. For Mars, at 1.5 AU, the solar flux is 2.25 ($= 1.5 \times 1.5$) times smaller than for Earth. In contrast, Venus, at 0.72 AU, receives solar flux almost twice as big as that of the Earth.

Mercury, at only 0.4 AU from the Sun, is lifeless. It's the smallest of the eight planets (about two-fifths the diameter of Earth) and has no liquid water and probably never did. Today, Mercury's barren surface sometimes reaches 430°C. If Mercury once had an atmosphere, it would have burned off when Mercury formed because of low gravity and intense sunlight.

The distance from the Sun explains why Venus and Mars are hostile to life, although that's not the whole story. The roughly 460°C surface of Venus is even hotter than Mercury gets, while Mars is an icy desert with an average temperature of −56°C. Venus is comparable in size to the Earth, but Mars is smaller—around half the diameter and one-ninth the mass of Earth. In fact, Mars's small size led to its poor habitability today, as we'll see later.

Although solar flux is one factor, the greenhouse effect on Mars and Venus also determines surface temperatures. The Martian atmosphere exerts only 0.006 bar surface pressure, compared to Earth's 1 bar. This wispy air is an arid mix of 95.3 per cent carbon dioxide, 2.7 per cent nitrogen, and minor gases. Recall that Earth's greenhouse effect is 33°C. Because of atmospheric thinness and dryness, Mars's greenhouse effect is only 7°C, which leaves the planet frozen. The two main gases in Venus's atmosphere, 96.5 per cent carbon dioxide and 3.5 per cent nitrogen, have similar proportions to those on Mars. But in stark contrast, Venus's atmosphere is massively thick and has a huge pressure of 93 bar at an average ground elevation. As a result, Venus's greenhouse effect is a whopping 507°C, i.e. 500°C bigger than on Mars (Table 2). The upshot is that neither planet's surface supports liquid water. In the thin air on Mars, a puddle would boil (or rapidly evaporate) and freeze at the same time, and water generally exists only as ice or vapour. Meanwhile on the scorching surface of Venus, liquid water is impossible.

Without liquid water, Venus is considered by most astrobiologists to be lifeless. A few speculate that acidophile microbes might live in its clouds of sulphuric acid particles. However, I doubt it. Apart from the lack of available water, atmospheric turbulence would pull microbes down into Venus's inferno or up to fatally desiccating heights.

Was Venus inhabited in the past?

Venus is still astrobiologically interesting because it may once have had oceans and life. It should have begun with plenty of water because the amounts of other volatiles are similar to Earth's. (Volatiles are substances that can become gases at prevailing planetary temperatures.) For example, if you took all of the Earth's carbonate rocks and turned them into carbon dioxide (CO_2), the Earth would have about ninety atmospheres' worth of CO_2, like Venus. If you extracted all the nitrogen in minerals on the Earth and added it to the Earth's atmosphere, you would get almost three atmospheres' worth of nitrogen gas, which again is similar to that in Venus's atmosphere. Because accretion of hydrated asteroids was the ultimate source of carbon and nitrogen on Venus and Earth, it's reasonable to infer that Venus also gained lots of water from them, as the Earth did.

Unfortunately Venus was doomed by its proximity to the Sun because of a *runaway greenhouse effect*. When baked by intense sunlight, the evaporation of water can make an atmosphere so moist that it becomes completely opaque to the infrared radiation emitted from the planet's surface. At this point, there's a limit to the cooling of the planet by emission of infrared radiation to space, which is set by the properties of water and a planet's gravity. Recent calculations suggest that this *runaway limit* is about 282 Watts per square metre for Earth and a few Watts less for Venus.

To understand the runaway limit, consider turning Earth into Venus. As mentioned earlier, the solar flux at the Earth's orbit is 1,366 Watts per square metre but the Earth reflects 30 per cent of this and so absorbs only 70 per cent. Then an additional reduction of 50 per cent comes from having only half of the Earth in daylight, and a further 50 per cent decrease accounts for glancing sunbeams on Earth's curved surface. Putting all these factors together, the Earth absorbs a net $(0.7 \times 0.5 \times 0.5) \times 1,366 = 240$ Watts per square metre of sunlight. When stable, the Earth emits the same amount of energy into space in the infrared and so keeps a constant global average temperature. But imagine if the Earth were moved to the orbit of Venus, where the absorbed sunlight per square metre would double from 240 to 480 Watts. The ocean would become like a hot bath and the steamy atmosphere would reach the runaway limit where only 282 Watts per square metre can radiate into space. With more energy per square metre coming in (480 Watts) than going out (282 Watts), the Earth would simply get hotter and hotter. The entire ocean would evaporate and surface rocks would melt. At that point, near-infrared light from a searing upper atmosphere would shine through to space, so that the incoming sunlight and outgoing radiation would come back into balance and the surface temperature would plateau around 1,200°C. This nasty process is what we think happened to Venus. Any Venusians would have been toast.

The runaway would also cause Venus's ocean to vanish. In the upper runaway atmosphere, ultraviolet light would split water vapour into hydrogen and oxygen. Hydrogen is so light that it would escape into space, dragging along some oxygen, while any oxygen left behind would oxidize hot rocks below. Eventually rocks would solidify, but by this time the atmosphere would be loaded with carbon dioxide and nitrogen released from the hot surface. The end result would be today's hellish Venus.

Before the runaway, life might have thrived in Venus's oceans. Recall that the early Sun was less luminous. So, a few hundred million years might have passed before the runaway commenced. Unfortunately, it might be tricky to prove if life ever existed. In the 1990s, radar on NASA's *Magellan* spacecraft mapped the Venusian surface. Older surfaces have more craters, so the density of impact craters was used to estimate the age of the surface as a fairly uniform 600–800 Ma. Venus's surface appears to have been repaved by lava all at once. A possible reason is that whereas Earth's internal heat is released continuously by creating new seafloor, on Venus internal heat might periodically build up until the interior becomes so hot that lava erupts everywhere. Venus might behave like this because unlike Earth it doesn't have water to lubricate plate tectonics.

Resurfacing would destroy any fossils. However, subtle geochemical traces might remain. To investigate if Venus had life, the best option would be to collect samples of Venusian rocks and return them to Earth in a future space mission.

Watery Mars: an abode for life?

Many have imagined Mars, unlike Venus, as a potential abode for life. Before the Space Age, basic parameters were known, such as Mars's 24.66 hour day, a year of 1.9 Earth years, and gravity that is 40 per cent of Earth's. Not much about the surface was certain until the first successful space mission, the flyby of NASA's *Mariner 4* in 1965, revealed a heavily cratered surface like the Moon. This put a damper on hopes for life. Then, in the early

1970s, NASA's *Mariner 9* orbiter photographed dried-up river valleys and extinct volcanoes, suggesting that Mars had once been quite Earth-like after all. But soon the pendulum swung back in the other direction. The *Viking* mission, consisting of two identical orbiters and landers, reached Mars in 1976, and failed to find life, as we discuss later.

After a hiatus in the 1980s, exploration was revived. *Mars Global Surveyor*, which orbited from 1997 to 2006, mapped Mars and imaged sedimentary layers that implied many geologic cycles of erosion and deposition. In the early 21st century, the *Mars Odyssey* and *Mars Express* orbiters, along with *Mars Reconnaissance Orbiter*, discovered areas of clay minerals and salts, which may have formed in liquid water. Twin *Mars Exploration Rovers* landed in 2004, and found fossilized ripples from past liquid water and sedimentary rocks, while in 2008, NASA's *Phoenix Lander* dug up subsurface ice in a polar region and measured soluble salts in the soil. Finally, in 2012, *Curiosity Rover* trundled towards a 5-kilometre-high mountain of sedimentary beds within a 150-kilometre-diameter crater named Gale. It found mudstones deposited from water that contain the SPONCH elements (Chapter 1) needed for life.

Nowadays, astrobiological interest in Mars concerns either biological traces from billions of years ago or microbial-like life that miserably endures underground. Pre-Space Age hopes of an Earth-like Mars today have been replaced with the reality of a cold, windswept, global desert with dust storms, daily dust devils, and no rainfall.

Mars's current surface is hostile to life for three reasons. First, while ice exists for sure, no liquid water has been unequivocally identified. The polar caps are water ice, topped with carbon dioxide 'dry ice' that grows when about 30 per cent of the atmosphere freezes at the winter pole. Also, above mid latitudes, ice-cemented soil or *permafrost* lies just beneath the surface. In the tropics, afternoon temperatures in the top centimetre of soil rise above freezing, but there, ice turns to vapour before melting temperatures are reached. A second problem is no ozone layer, allowing harmful ultraviolet sunlight to reach the surface. Third, chemical reactions in the atmosphere make hydrogen peroxide—the same chemical used in hair bleach. Hydrogen peroxide molecules settle to the surface, where they can destroy organics.

While the surface is unpromising, geothermal heat underground might allow liquid water and life to exist. In fact, from 2004, reports of atmospheric methane averaging ten parts per billion by volume led to excitement that subterranean methanogens might be present. Sunlight reflected from Mars gathered by telescopes and *Mars Express* appeared to show absorption by atmospheric methane. However, the methane signal was barely distinguishable and some sceptical scientists (including me) doubted whether methane was really present. Subsequently, the *Curiosity Rover* has failed to detect methane down to levels of one part per billion.

In discussing past life, astrobiologists refer to the Martian geological timescale, which is divided into aeons called the *Pre-Noachian* (before 4.1 Ga), *Noachian* (4.1 to about 3.7 Ga), *Hesperian* (3.7 to 3.0 Ga), and *Amazonian* (since 3.0 Ga). Surfaces on Mars are placed into each aeon according to impact craters. Older surfaces have accumulated bigger and more numerous craters. In fact, the dates bounding each aeon actually come from the Moon. The ages of rocks brought back by the Apollo astronauts are known from radioisotopes and these correlate lunar cratering densities with time. Astronomical calculations that account for more impactors on Mars than the Moon allow the lunar correlation to be extended to Mars.

Evidence that liquid water used to be present suggests that Mars was once more habitable than today. Images show *fluvial* (stream-related) features in the landscape, including gullies, dried-up river valleys, deltas, and enormous channels. Also, the soil and rocks contain minerals that form in the presence of liquid water.

Gullies are incisions of tens to hundreds of metres length on the walls of craters and mesas between 30 and 70° latitudes in both hemispheres. Because gullies lack superimposed craters and sometimes flow over sand dunes, they must be very recent. Initially, they were thought to form when ice melted. However, images show them forming when carbon dioxide frost vapourizes, presumably releasing a dry flow of soil and rocks.

Far more ancient features are *valley networks*, which are dried-up river-like depressions that spread out in tree-like branches with tributaries (Fig. 8). Most incise heavily cratered Noachian terrain and they're 1–4 km wide and 50–300 m in depth. The density of tributaries is far less than for most terrestrial rivers, but sometimes enough to infer the drainage of rain or meltwaters. In other cases, valleys with stubby tributaries were probably formed by *sapping*, when underground (melt)water caused erosion and collapse of the overlying ground. A few valleys end in deltas.

The Noachian landscape consists of craters with degraded rims and shallow floors. What caused this erosion is unclear. Valley networks are incised on top and so weren't responsible. Once the Hesperian started at 3.7 Ga, erosion rates dropped dramatically and valley networks became rare.

Towards the end of the Hesperian (around 3 Ga), *outflow channels* appeared (Fig. 8). Channels form from fluid flow confined between banks, lacking tributaries and emerging from a single source, unlike river valleys. The outflow channels are huge: 10–400 km width, up to 1,000 kilometres or so long, and up to several kilometres deep. Most begin in *chaotic terrain* where the ground collapsed, sometimes in canyons or chasms where there

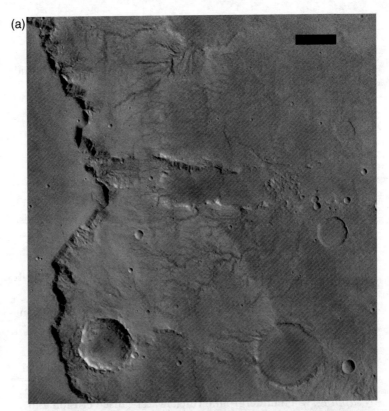
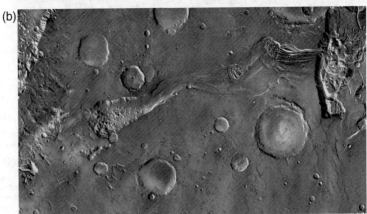

8. a) Valley networks on Mars. On the left is the eroded rim of 456-km-diameter Huygens Crater. The image is centred at 14°S, 61°E, north upwards. Scale bar = 20 km; b) Outflow channel Ravi Vallis (0.5°S, 318°E) about 205 km long

are mountain-sized heaps of sulphate salts, which might be evaporation residues from salty water.

The leading explanation for outflow channels is that they formed from floodwaters when underground ice melted or aquifers burst. However, they require 10–100 times more flow than the best-known terrestrial analogue: flood-carved land across eastern Washington State, USA, which early settlers called *scablands*. The scablands formed at the end of the last ice age when ice damming of a large lake periodically ruptured.

It's not easy to explain the water needed to erode Mars's outflow channels. Estimates suggest the equivalent of a global ocean several hundred metres deep, which is far more water than exists as ice today. Since the channels flow to the northern lowlands, some scientists speculate that an ocean formed there. In contrast, a minority thinks that the outflow channels were not carved by water but by lavas. Chemistry suggests that Martian lavas should have been runny, turbulent, and erosive.

Apart from ancient valley networks and outflow channels (if water eroded), minerals provide other evidence for a wetter early Mars. Everyone knows how lettering on old gravestones disappears because of chemical reactions with water. Such reactions change or dissolve minerals in *chemical weathering*. Much of the Martian surface, like that on Venus and the Earth's seafloor, is made of basalt, a dark-coloured igneous rock rich in iron and magnesium silicate minerals. When basalt is chemically weathered, *alteration minerals* are produced, such as clays. So the presence of alteration minerals means that liquid water was present, sometimes with a specific pH. For example, alkaline waters tend to produce clay minerals from basalt.

Hydrous (water-containing) alteration minerals have been identified from analysis of infrared radiation emitted and reflected by Mars's surface. But only about 3 per cent of Noachian surfaces have hydrous clays and carbonates. Sulphate minerals dominate late Noachian or Hesperian areas, while reddish, dry iron oxides are common on younger Amazonian surfaces. This pattern might imply three environmental epochs. In the first epoch, alkaline or neutral pH waters weathered basalt and made clays. During the second, sulphuric acid was derived from volcanic sulphur gases and made the sulphates. The third epoch continues today with a cold, dry environment and rust-coloured surfaces.

One of the twin *Mars Exploration Rovers*, named *Opportunity*, actually landed near Noachian sulphates. Millimetre-sized spheres of the iron oxide hematite (Fe_2O_3) were embedded in sulphate layers, like blueberries in a muffin. The hematite precipitated from minerals carried in water percolating underground about 3.7 billion years ago. Also, ankle-deep water appears to have ponded on the surface, leaving behind ripples in the sediment. The other Rover, named *Spirit*, found evidence of ancient hot springs on the opposite side of the planet.

The early atmosphere and climate of Mars

Evidence of liquid water leaves us wondering whether the climate on early Mars was warm and wet. Scientists disagree and fall into two camps. One group argues that Mars had a warm, wet climate for tens or hundreds of millions of years. The other maintains that transient melting of ice in a cold climate could account for what is seen.

The first scenario is obviously more favourable for life. But unfortunately, no one has yet explained how early Mars was kept warm for millions of years. Some 3.7 billion years ago, the Sun was 25 per cent fainter. A greenhouse effect of about 80°C would have been needed to keep early Mars just above freezing, compared to Earth's modern 33°C greenhouse. It's generally thought that any atmospheric hydrogen should have escaped into space rapidly when Mars formed, leaving the atmosphere oxidized and full of carbon dioxide and nitrogen. Mars couldn't ever have had an extremely thick Venus-like atmosphere because carbon dioxide would condense into ice and form clouds at Mars's distance from the Sun. A thick carbon dioxide atmosphere is also good at scattering sunlight back to space, which cools the surface. So carbon dioxide can't provide an early warm climate. An alternative suggestion is that volcanic sulphur dioxide was the key greenhouse gas. However, sulphur dioxide dissolves in rain and would be flushed from the atmosphere if Mars became wet. Also, atmospheric reactions at high altitudes make a fine suspension of sulphate particles from sulphur dioxide, which reflects sunlight and cools the planet. Such cooling happened on Earth during 1991–3 as a result of volcanic emissions from Mt Pinatubo in the Philippines.

The other camp proposes many mechanisms that allow liquid water in a cold climate. They note that impacts would have vaporized ice into steam, which, in turn, would have produced rainfall that eroded river valleys. Also, erosion might have been produced by local snowmelt as a response to past fortuitous combinations of Mars's axial tilt and orbital shape. These characteristics are perturbed over time by the gravitational influence of other planets. The tilt of a planet's axis with respect to the planet's orbital plane, e.g. 23.5° for Earth, causes the seasons because one hemisphere gets more sunlight at one point in the orbit than the other. Unlike Earth, Mars has no big moon to stabilize its axis (its two tiny moons have negligible effect), so Mars's tilt has varied between 0° and 80° over the last 4.5 billion years. At high tilts, summertime polar ice faces the Sun and vapourizes. Air currents transport the vapour to the cold tropics where it snows. Sunlight during other seasons or at lower tilts could then produce meltwaters and fluvial erosion. Moderate tilts in the last few million years might explain relict midlatitude patches of dust that were once ice-cemented. Finally, salty water on Mars can remain liquid far below 0°C. On Earth, we spread sodium chloride on icy roads in winter because it melts ice down to −21°C. Another salt, perchlorate, which was detected in Martian soil by the *Phoenix* lander in magnesium or calcium form, can depress the freezing point of water below −60°C.

Whatever the truth about early Mars, Mars's small size ultimately spoiled its habitability. Because small objects cool faster than large ones, internal heat was lost rapidly, so that widespread volcanism ceased long ago. Without volcanism, Mars couldn't recycle carbon dioxide, leaving the gas to be converted into carbonates, which are present but not in the abundance expected if a thick carbon dioxide atmosphere had all been transformed. So, in addition, part of the early Martian atmosphere was probably blasted away to space by large comet and asteroid impacts in so-called *impact erosion*. The atmosphere was vulnerable because of Mars's low gravity. Along with more gradual escape of gases to space later, Mars was ultimately left with its thin atmosphere.

Looking for life on Mars

We still don't know if Mars has life or ever had life, despite attempts to find it. The *Viking* landers tried to detect life in 1976, and, since the 1990s, scientists have looked for signs of past life in Martian meteorites.

Each *Viking* lander had three experiments to detect metabolism and a fourth to find organic molecules. The first, the *carbon assimilation experiment*, examined if Martian microbes obtained carbon from air. Martian soil (sometimes mixed with water) was exposed to carbon dioxide (CO_2) and carbon monoxide (CO) gases brought from Earth, with carbon-14, a radioactive isotope, in each gas. Afterwards, the soil was found to have incorporated carbon-14. When a 'control experiment' was done where the soil was first sterilized at 160°C, the soil still took up carbon-14, which suggested that inorganic chemistry was responsible, not Martian microbes. A second test was the *gas exchange experiment*, which monitored Martian soil and a solution of organics brought from

Earth to see if gases were generated from metabolism. Strangely, O_2 was released, but it was also released from sterilized soil. Evidently, chemicals in the soil decomposed into O_2 with water or heat. The third investigation, the *labelled release experiment*, added organics containing carbon-14 to soil. If Martians metabolized the organics, they would give off carbon-14-containing CO_2. Radioactive gas was emitted, while sterilized soil released no radioactive gas. At face value, this was a positive detection of life. But most scientists think that soil oxidants reacted with organics to make CO_2 and were inactivated by heat. A reason to prefer this explanation came from the fourth experiment, in which a sort of electronic nose, called a *gas chromatograph mass spectrometer*, identified molecules wafting off heated soil. It found no organic material in the soil to a detection limit of a few parts per billion.

Overall, the moral is that before looking for extraterrestrials, you need to understand the inorganic chemistry of the environment to avoid false positives. In fact, Harold 'Chuck' Klein (1921–2001), who led the *Viking* biology experiments, told me that he wanted to do this when the *Viking* mission was conceived, but managers at NASA who controlled finances instead insisted on looking for life directly.

While the Viking results were being pondered in the 1980s, a remarkable discovery was made: there were already rocks from Mars on Earth—Martian meteorites! Impacts knock rocks off Mars and some of these fall on the Earth. In fact, about 50 kg lands every year, mostly in the ocean. Gas sealed within some meteorites during minor melting associated with impact ejection proves a Martian origin because it matches the atmosphere measured by the *Viking* landers. Other Martian meteorites without trapped gas have a triple oxygen isotope (^{16}O, ^{17}O, ^{18}O) composition in their silicate minerals that is unique to all Martian meteorites and identifies them like a fingerprint. By 2013, sixty-seven Mars meteorites were known but the list keeps growing.

In 1996, possible signs of past life in ALH84001 were reported. This Martian meteorite was the first one (the '001') collected in the Allan Hills, Antarctica (the 'ALH'), in a 1984 expedition (the '84'). ALH84001 crystallized as an igneous rock on Mars at 4.1 Ga. Inside the rock are some carbonate globules about 0.1 mm across that formed at 4.0–3.9 Ga, and within them are four possible traces of life: alleged microfossils; carbonates said to be precipitated by microbes; traces of organics called polycyclic aromatic hydrocarbons (PAHs), which are made of hexagonal rings of carbon atoms; and crystals of the mineral magnetite (Fe_3O_4) said to be similar to those within certain bacteria. On Earth, magnetotactic bacteria make magnetite crystals inside their cells with shapes that evolution has honed into magnets. The bacteria use magnetite compasses to move along the up–down component of the Earth's magnetic field in order to find a boundary between a lower oxygen-poor zone below and an upper oxygenated zone, which optimizes their metabolism.

In subsequent years, research has cast doubt on all four claims. The alleged microfossils are rod-shaped structures that merely *look like* microbes. Scientists have since found inorganic mineral surfaces with similar shapes. Also, the structures in ALH84001 are about ten times smaller than terrestrial microbes and probably beyond the minimum volume needed for essential biochemistry. An evaporating fluid can produce the carbonates in the globules and so there's little reason to invoke biogenic salts, proposed as the second feature. Regarding the third line of evidence, analyses showed that most PAHs got into the meteorite while it was sitting in Antarctica for 13,000 years. PAHs are ubiquitous atmospheric pollutants produced from burnt organic material. Meteorites are dark and absorb sunlight well, so they can sit in puddles in Antarctic ice during summertime, allowing water and chemicals to infiltrate cracks. If there are Martian PAHs in the interior of ALH84001, these could have been made inorganically. The rock was heated by impacts that occurred before the ejection impact. Heating should have released carbon-bearing gas from the carbonates, which can react with water to make organic matter without life. The fourth argument concerned magnetite. It turns out that only a tiny fraction of magnetites in ALH84001 have biogenic-like shapes. Some scientists argue that magnetite of many shapes, including bacteria-like ones, formed during heat shocks prior to ejection when iron carbonate decomposed into magnetite.

The controversy about ALH84001 shows how difficult it is to prove that microscopic life exists in old rocks. But since most terrestrial rocks lack fossils, absence of life in ALH84001 doesn't mean that life on Mars didn't exist. We need to keep looking.

Ceres: a serious candidate for habitability?

Beyond Mars and within Jupiter's orbit is an asteroid belt of millions of small rocky bodies, including the largest, Ceres, some 950 km in diameter, at 2.8 AU from the Sun. Ceres's surface contains clay minerals and water ice. As on Mars, clays suggest past liquid water. A leading model for Ceres's internal structure is a rocky core that's surrounded by a 100-km-thick shell of ice. Just above the core, there may be a subsurface ocean. This could be very salty, allowing its persistence at sub-zero temperatures, and perhaps it once flowed to the surface.

Could life exist in hydrothermal vents at the bottom of Ceres's ocean? Ceres is so small that there's probably little internal heat available today, so any biomass would be extremely meagre. But perhaps in the past, Ceres was more habitable. NASA's *Dawn* mission will arrive at Ceres in 2015. Spectra of infrared emission and reflected sunlight should give us more detail about the surface composition and habitability.

The icy Galilean moons of Jupiter

Beyond the asteroid belt is Jupiter, which is probably not a realistic target for astrobiology for reasons given in Chapter 5. But

9. The Galilean moons of Jupiter: Io, Europa, Ganymede, and Callisto

Jupiter has sixty-seven moons and three might be habitable. Galileo discovered the four largest moons in 1610, which are Io, Europa, Ganymede, and Callisto, going outwards from Jupiter (Fig. 9). Io and Europa are similar in size to the Moon, while Ganymede and Callisto have proportions comparable to Mercury. The outer three satellites have rocky interiors covered with ice, while Io's exterior is just rocky. Io, Europa, and Ganymede probably also have iron cores.

Io is the most volcanic body in the Solar System, with volcanoes spewing sulphur dioxide, which freezes onto Io's surface. But Io is devoid of liquid water and surely lifeless. Io is so volcanic because gravitational forces from Jupiter vary as Io goes round in an elliptical orbit, squeezing Io like a stress ball. This *tidal heating* of Io is caused by friction when solid material moves up and down, similar to the way that water sloshes in Earth's oceanic tides. Io's orbit is forced to be elliptical because of periodic gravitational prodding from Europa and Ganymede. The times that Ganymede, Europa, and Io take to make orbits have a ratio of 4:2:1, which causes the moons to line up cyclically and nudge each other gravitationally. This relationship is the called the *Laplace resonance*, after Pierre-Simon Laplace, the French mathematician mentioned in Chapter 2.

Europa also has an orbit that's forced to be elliptical and so it has tidal heating too. The heating is smaller than that of Io because Europa is further from Jupiter. The average temperature of

Europa's icy surface is −155°C, but tidal heat can maintain liquid water deep under the ice. Impact cratering densities suggest a surface age of only 20–180 Ma consistent with slush from below repaving the surface. Images show *chaotic terrain*, i.e. disrupted surfaces with blocks that have shifted, which suggest that subsurface ice might convect. A global network of stripes and ridges includes bands where the surface has ripped apart and ice appears to have been extruded. Magnesium sulphate salt is present on the surface, perhaps originating from waters within, while unknown materials cause reddish streaks.

The key evidence for Europa's subsurface ocean comes from magnetic measurements by NASA's *Galileo* spacecraft, which orbited Jupiter from 1995 to 2003. Unlike Earth or Jupiter, Europa's interior doesn't create its own, *intrinsic* magnetic field. But in passing through Jupiter's large magnetic field, electrical currents are induced inside Europa. In turn, these currents generate a weak and varying *induced magnetic field*. The strength of this field requires that an electrically conducting fluid exist within 200 km of Europa's surface. The most likely explanation is a salty ocean up to twice the volume of Earth's ocean. The largest craters suggest that the ice cover above the ocean is at least 25 km thick. However, some terrain that pokes up might be produced by upwelling ice that melts a few kilometres below the surface and then refreezes. So shallow lens-shaped lakes might exist.

Besides liquid water, the possibility of life depends on energy sources, interfaces, and the availability of SPONCH elements. Europa probably has a good supply of biogenic elements if it was made of material similar to carbon-rich meteorites or comets. Moreover, heat-releasing reactions of water and a rock seafloor could supply energy as well as nutrients. *Radiogenic heat*, which is produced by the decay of radioactive isotopes within the rock (such as potassium, uranium, and thorium), might produce seafloor vents that supply carbon dioxide and hydrogen for chemoautotrophs.

The availability of oxygen in the ocean would also permit biological oxidation of iron and hydrogen on the seafloor. There are two small sources of oxygen. Charged particles trapped in Jupiter's magnetic field slam into Europa's icy surface and break water molecules, releasing oxygen. If the ice churns, some of the oxygen could be carried down to the ocean. Other oxygen comes from water molecules split by radiation from radioactive elements. Estimates of Europa's biomass vary from around a thousand to a billion times less than on Earth. Some optimists even speculate that there might be enough oxygen for animal-like creatures!

Measurements of induced magnetic fields on Ganymede and Callisto also imply subsurface oceans but deeper, below roughly 200 km and 300 km depth, respectively. Ganymede's average surface age is about 0.5 Ga, so its ocean has probably not gushed to the surface recently. Because Ganymede is Jupiter's largest moon, its interior supports high-pressure forms of ice that would lie beneath the ocean. Lack of a rock–water interface might be less favourable for life than on Europa. The lower tidal heating on Ganymede also implies a smaller biomass.

Inference of a subsurface ocean on Callisto is surprising because Callisto lacks tidal heating. The warmth must be radiogenic and the ocean might be loaded with salts that lower the freezing point. Callisto's surface age is 4 Ga, so the prospect of finding evidence for oceanic life there might be slim. Also, with less energy available, any biomass should be even smaller than that on Ganymede.

Enceladus and Titan: icy moons of Saturn

Saturn and its sixty-two moons lie at about twice the distance of Jupiter from the Sun. The largest moons, discovered before the Space Age, are (going outwards): Mimas, Enceladus, Tethys, Dione, Rhea, Titan, Hyperion, Iapetus, and Phoebe.

Enceladus is special because it has active geology. It's Saturn's sixth largest moon, but its 500 km diameter could almost fit within Great Britain. Enceladus orbits Saturn twice for every orbit of Dione. As a result, periodic gravitational nudges from Dione force Enceladus's orbit to be elliptical, resulting in varying gravitational forces from Saturn that flex and heat Enceladus. For reasons that are not fully understood, the heating is concentrated under Enceladus's south pole. There, icy particles and gas spray out of parallel fractures dubbed *tiger stripes*, which are about 130 km long, 2 km wide, and flanked by ridges. The jets contain traces of methane, ammonia, and organic compounds, along with salt. In fact, the jets supply particles to the 'E-ring' of Saturn within which Enceladus orbits.

Enceladus has a rocky core, an icy shell, and an underground sea beneath the area of the jets. The jackpot combination of organic molecules, energy, and liquid water implies that life might exist inside Enceladus. Such life could be chemoautotrophic, feeding off hydrogen produced by water–rock reactions or hydrogen and oxygen from water that's split by radioactivity. If life does exist there, methane or organic matter in the jets could be biological.

Titan, the largest moon of Saturn, is also of astrobiological interest. It's similar in size to Ganymede and Mercury and is the only satellite in the Solar System to have a thick atmosphere. In fact, the air pressure at Titan's surface is 1.5 bar, which is 50 per cent larger than that at the Earth's surface. The atmosphere of 95 per cent nitrogen and 5 per cent methane provides a 10°C greenhouse effect, but the sunlight at Saturn's distance is one hundred times less intense than at Earth, so Titan's surface is incredibly cold at −179°C.

Titan's atmosphere contains a smoggy haze of hydrocarbons. At high altitude, ultraviolet sunlight breaks up methane molecules (CH_4) and subsequent reactions build up hydrocarbons including ethane (C_2H_6), acetylene (C_2H_2), propane (C_3H_8), benzene (C_6H_6), and reddish-brown particles containing polyaromatic hydrocarbons. The particles are called *tholins* from the Greek *tholos* for 'muddy.'

The products derived from methane sediment out of the atmosphere. In fact, about 20 per cent of Titan's surface is covered in tropical dunes that are made of sand-sized particles, which are at least coated with organics if not made of them. At the poles, over 400 lakes are mixtures of liquid propane, ethane, and methane, with extraordinary beaches made of particles of benzene and acetylene.

Methane on Titan behaves like water on Earth and forms clouds, rain, and rivers. Most of our knowledge about Titan comes from the *Cassini–Huygens* mission, which arrived in 2004. *Huygens* landed on Titan in 2005, while *Cassini* is a Saturn orbiter. Near the landing site of *Huygens*, channels were eroded by liquid hydrocarbons (Fig. 10). At the landing site itself, cobbles (presumably made of water or carbon dioxide ice) have been rounded as if by transport in rivers of liquid methane. Methane rain should be very infrequent, but when it rains, it pours.

Because methane is destroyed by sunlight, Titan's atmosphere would run out of methane in about 30–100 million years if it were not resupplied. How methane is replenished is a mystery. The leading hypothesis is that methane leaks out of Titan's interior, which suggests geological movement inside Titan.

Titan flexes too much to be entirely solid, which is evidence that Titan has a subsurface ocean. The eccentricity of Titan's orbit ensures that Titan is squeezed by Saturn's gravity. The ocean exists below an icy crust of less than 100 km thickness. But as with Ganymede, the seafloor should be dense ice rather than rock because Titan's mass allows such high-pressure forms of ice.

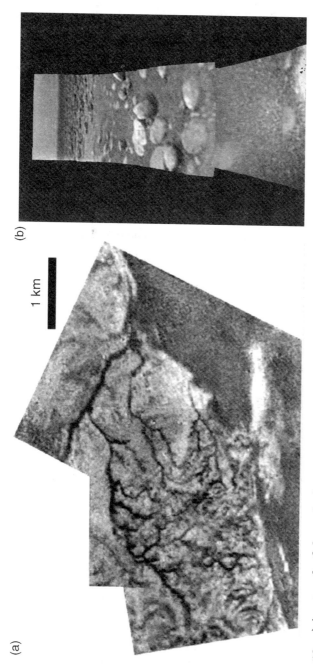

10. a) A network of channels that appear to flow into a plain near the Huygens landing site; b) Image of the surface at the Huygens landing site. Stones in the foreground are 10–15 cm in size and sit on a darker, fine-grained substrate

Two types of life might exist on Titan: Earth-like life in the subsurface ocean; and weird life in the hydrocarbon lakes. Regarding the first possibility, when Titan formed, the heat released from the capture of smaller bodies should have created liquid water on the surface temporarily. Perhaps water-based life evolved and survived as an ocean retreated underground.

Weird life would be limited in biochemical complexity because oxygen, which is scarce, is required for sugars, amino acids, and nucleotides. Titan captures some oxygen-containing molecules from space, but the supply is tiny. Hypothetical life in cold hydrocarbon solvents might use acetylene for energy. On Earth, acetylene burns with oxygen in welding torches. But Titanians could metabolize acetylene with hydrogen from Titan's air. The chemical reaction,

$$C_2H_2 (\text{acetylene}) + 3H_2 (\text{hydrogen}) = 2CH_4 (\text{methane})$$

produces energy, but the idea is speculative. The main problem for weird life on Titan, I think, is that liquid hydrocarbons are poor at dissolving large molecules, which are essential for genomes. Larger molecules are less soluble than smaller ones in liquid hydrocarbons. Furthermore, solubility declines at lower temperatures.

Some exoplanets might have Titan-like liquid hydrocarbons. The most common stars are red dwarfs, which are smaller and colder than the Sun. Exoplanets near 1 AU distance around such dwarfs would have surface temperatures in the range of liquid hydrocarbons, not liquid water. Thus, if we were to discover weird life on Titan, we might imagine a universe teeming with life utterly different from our own. Ironically, we would then be the weird life.

Triton: a captured Kuiper Belt object around Neptune

At and beyond the orbit of Neptune at 30 AU, objects are covered in nitrogen ice (N_2). Triton, the largest of thirteen moons of

Neptune, and Pluto are both really *Kuiper Belt objects* (KBOs). The *Kuiper Belt* is a region of icy bodies within the plane of the Solar System at 30–50 AU left over from Solar System formation. There are also KBOs scattered out to 1,000 AU. Triton was captured by Neptune because two features of its orbit suggest so: Triton orbits in the opposite direction to Neptune's rotation, and its orbital plane is tipped up 157° with respect to Neptune's equator.

Triton's surface is tremendously cold, around −235°C, because it reflects 85 per cent of sunlight. An extremely thin nitrogen atmosphere exerts 20 millionths of a bar surface pressure. This 'air' is simply the nitrogen vapour that will sit over nitrogen ice at Triton's prevailing temperature. There's also a little methane (about 0.03 per cent of the nitrogen concentration), which is destroyed by ultraviolet sunlight, creating a thin smog of hydrocarbon particles. If methane destruction has been operating at the same rate for 4.5 billion years, about a metre's depth of organic material should have accumulated. However, mobile frosts would cover this up. A reddish tint to some of the ice might be the organics. In fact, sunlight sometimes vaporizes nitrogen ice into geyser-like plumes, which carry dark particles that are perhaps organic. Apart from nitrogen ice, water and carbon dioxide ice make up some of the surface.

Triton is geologically active because crater counts suggest that resurfacing is only 10 Ma and there are icy structures like vents, fissures, and lavas. We call these *cryovolcanic*, meaning a form of volcanism in which slushy ice comprises the equivalent of lava and magma. Triton's density suggests a large rocky core that supplies sufficient radiogenic heat for cryovolcanism and potentially a subsurface ammonia-water ocean.

Early Triton may have been more habitable. Just after capture, Triton's orbit would have been highly elliptical, producing huge tidal heating from Neptune that should have melted an extensive subsurface ocean. Tides tend to circularize orbits and today Triton's orbit is close to circular. Thus, after capture, the ice shell gradually thickened as tidal heating declined. Today, the ocean might be below hundreds of kilometres of ice, and would be underlain by high-pressure forms of ice. But although oceanic life might be deeply hidden, the presence of cryovolcanism suggests that if we sample organic matter on the surface, there's a chance of finding traces of life.

Does Pluto have an underworld ocean?

Pluto, named after the god of the underworld, has a surface mainly of nitrogen ice, and a thin nitrogen atmosphere of 8–15 microbar surface pressure. There's also a little atmospheric methane that sunlight destroys just as on Triton and Titan. Again, this produces hydrocarbon particles, which settle out, possibly accounting for dark areas on Pluto's surface.

A collision between Pluto and another KBO is believed to have created Charon, the largest of Pluto's five moons, in an event analogous to the impact that formed the Earth's Moon. Thus, Pluto is really a double KBO: Charon is half Pluto's size and one-ninth its mass. Every 6.4 days they orbit around a point that lies between them. After the Charon-forming impact, tidal heating should have melted a subsurface ocean on Pluto. Pluto probably has a rocky core that might supply enough radiogenic heat to maintain a subsurface ocean of ammonia-rich water today.

Organisms in Pluto's ocean would be limited by energy, so the biomass would be far smaller than on Europa. Any ocean is also probably at depths exceeding 350 km, and difficult to access in a future space mission, but life might be there.

Our consideration of Pluto shows that life might exist in remarkably unlikely places, and my list in Table 1 may be too conservative. Many of the cold outer Solar System objects used to be dismissed as potential habitats. But perceptions have shifted. It's possible that there are subsurface ammonia-rich oceans on Saturn's moon Rhea, Uranus's largest moons Titania and Oberon, and perhaps some large KBOs, such as Sedna, which is similar in size to Pluto yet 100 AU from the Sun. Given that so many objects in the Solar System are potential refuges for life, billions of similar possibilities surely exist throughout our galaxy.

Chapter 7
Far-off worlds, distant suns

The hunt for exoplanets

Beyond the Solar System, astronomers have discovered over 3,400 exoplanets, including candidates and confirmed bodies. It's easier to detect large ones, so nearly all are bigger than Earth and some are rather exotic. *Hot Jupiters* are Jupiter-size planets within 0.5 AU of their parent stars, some orbiting in only a few days. Then there are planets that sound like Earth on steroids, the *Super-Earths*. These are up to ten times the mass of our planet. Although harder to find, it's clearly just a matter of time before many Earth-sized exoplanets become known. How many will be habitable or even inhabited?

Of course, before identifying a *habitable* exoplanet, you have to find exoplanets. With the vast distances involved, the search is difficult but nonetheless astronomers have developed two classes of methods. The first, *indirect detection*, looks for stellar properties, such as position or brightness, which are affected by the presence of unseen planets. The second is *direct detection* of a planet with an image or a spectrum of its light.

The four indirect detection methods are: astrometry, stellar Doppler shift, transits, and gravitational microlensing. The key to the first two is that a planet and star orbit a common centre of mass. For example, if a planet and star had exactly the same mass, they would orbit a point halfway between them. In reality, a planet is smaller than a star, so the centre of mass is closer to the star and perhaps inside it. But in all cases, the star will follow a little orbit around the centre of mass and 'wobble' even when its planets can't be seen. In our own Solar System, Jupiter's twelve-year orbit causes twelve-year wobbles of the Sun's position. Saturn adds in another smaller twenty-nine-year wobble, corresponding to its twenty-nine-year orbit around the Sun.

Astrometry measures the motion of a star in the sky using telescopes. It's sensitive to big planets far away from their star so it could be used to find planetary systems similar to our own. But you have to wait many years or decades to track the effects of planets far from their star.

The second technique, the *Doppler shift* or *radial velocity* method, relies on the fact that when the light of a star is split into all its colours, the spectrum has dark bands like a bar code. The elements in a stellar atmosphere absorb photons coming from the star's interior, which cause the dark lines. If the star moves toward or away from the Earth, the lines shift to higher (bluer) or lower (redder) frequencies, respectively. Everyone has experienced the *Doppler effect* in sound. When a wailing police car approaches, the sound waves are bunched into a high-frequency squeal, but after the car passes by, they become a low-frequency drone. A similar frequency shift occurs with light. A rhythmic red shift and blue shift of lines in a stellar spectrum shows that the star is wobbling and has a planet around it. The size of the shift indicates the mass of a planet, while the pacing gives the time for the planet to complete an orbit.

In 1995, Didier Queloz (then a student) and his mentor, Michel Mayor, from the University of Geneva, detected the first planet around a Sun-like star using Doppler shift. It was a planet of at least half a Jupiter mass orbiting a star in the constellation of

Pegasus every four days. It was a total surprise. No one expected a giant planet so close (0.05 AU) to its parent star because giant planets shouldn't form there. Now we understand that some extrasolar planets undergo planetary migration early in their history (Chapter 3) and we end up seeing what survived the jostling. In 2012, Doppler data suggested a possible Earth-mass planet just 4.3 light years away, orbiting the star Alpha-Centauri B. The planet, if present, lies 0.04 AU from Alpha-Centauri B, which is so close that its surface is probably molten rock.

One subtlety with the Doppler shift technique is how we view the planetary system. If the orbit is face-on (described as an inclination of zero degrees from a plane perpendicular to the line of sight), there's no to-and-fro Doppler shift. The more edge-on the orbit, the greater the Doppler shift. It becomes maximal at an edge-on inclination of 90 degrees. There's often no way of knowing the inclination, so a measured Doppler shift might be smaller than the ideal edge-on case, allowing us only to infer a *minimum* planetary mass. The Doppler technique is most sensitive to big planets close to their star, so that's why most of the exoplanets initially detected were hot Jupiters. But we now know from the transit method that hot Jupiters are actually only a tiny minority of exoplanets, around 0.5 to 1 per cent.

The *transit* method measures the decrease in starlight when a planet crosses the face of a star, which can happen if we're lucky enough to view an exoplanet system virtually edge-on. Such geometry is statistically rare, but there are so many stars that if you gaze widely and long enough, transits will be seen. From the dip in starlight, it's possible to determine a planet's diameter if the star's size can be estimated. In turn, the planet's orbital period and distance from its star can be calculated from the cycling of the dimming. NASA's *Kepler* mission is a telescope that operated from 2009 to 2013 from an orbit around the Sun staring continuously at 145,000 main sequence stars in the constellation of Cygnus (the Swan) to look for transits where most of the stars are

500–3,000 light years away. *Kepler* has found over 3,200 exoplanet candidates, which are confirmed by checking for periodic dimming or using another detection technique, such as Doppler shift. In 2017, the Transiting Exoplanet Survey Satellite (TESS) telescope will begin examining two million stars for transits.

The fourth indirect method is *microlensing*. When an object passes between a distant star and us, Einstein's Theory of General Relativity predicts the bending of light by the object's gravitational field. The foreground object effectively acts as a lens, focusing the light and making the distant star appear gradually brighter. But if a planet orbits a lens star, the background star can brighten more than once. This sensitive technique can find Earth-mass planets at orbital distances of 1.5–4 AU around a star. Unfortunately, once the alignment has happened, it's virtually impossible to follow up with more detailed measurements because the objects are typically very distant. However, microlensing can gather statistics. Large planets drifting alone in interstellar space can also act as lens objects, and one result is that there appears to be many unbound Jupiter-mass planets floating between the stars. These lonely worlds were presumably ejected from their extrasolar systems.

For astrobiology, the most important techniques are *direct detections* that capture light from a planet. Direct detection is challenging because a planet is a dim body close to a vastly brighter star. Nonetheless, telescopes in space, such as the Hubble Space Telescope, and extremely large telescopes on the ground have accomplished direct detection using a *coronograph*, which is a mask to block the starlight. The term 'coronograph' comes from techniques that were originally used to block out sunlight to study the Sun's corona, which is the wispy halo seen in a solar eclipse. Ground-based telescopes also use *adaptive optics*, which are procedures to sense and correct the distortions caused by the shimmering of the Earth's atmosphere.

A space-based telescope has the advantage of avoiding the blurs from the Earth's atmosphere, but there's the question of which part of the spectrum to examine. In general, planets emit mostly infrared radiation and reflect visible starlight. If we looked at our Solar System from afar, the Sun, being hot and big, would outshine the Earth by a factor of about ten million in the infrared and ten billion in visible light. So looking for a distant Earth in the infrared gives a thousand times better contrast than in visible light. Unfortunately, light also spreads and blurs (diffracts) when it encounters any object such as a telescope and this is worse with infrared light than with visible. Fortunately, a technique called interferometry can cancel unwanted light. Noise-cancelling headphones use the same principle to silence undesirable sound waves.

Light waves have crests and troughs like water waves. *Nulling interferometry* uses more than one telescope mirror to precisely align light waves arriving from a point in the sky so that wave crests cancel troughs and light is turned into darkness. In this way, starlight can be 'nulled' to see planets. Both NASA and the European Space Agency have studied space telescopes, called *Terrestrial Planet Finder* and *Darwin*, respectively, which use interferometry to capture images and spectra of exoplanets around nearby Sun-like stars.

The results of exoplanet surveys to date generally show that the number of exoplanets increases at lower masses, so rocky, Earth-sized planets should be common. Density estimates for some exoplanets also allow assessment of composition. Density has been inferred using mass from the Doppler method and size from transit methods. Also, if a transiting Earth-sized planet has at least one Neptune-like sibling, a method called *transit timing variations* can provide the mass of planets and thus density. Even if the larger planet doesn't transit its parent star because of its orbital inclination, it can cause slight variations in the smaller planet's transit times that allow planetary masses to be calculated. With density measurements, there are increasing clues about which exoplanets are made of gas, rock, water, or some mixture.

The habitable zone

When we were considering life in the Solar System, we concentrated on bodies with liquid water because we're sure that's a requirement for Earth-like life, and the same idea applies to exoplanets. Whereas dry planets with subsurface oceans are worth investigating with space probes within our Solar System, they generally wouldn't be targets for exoplanet astrobiology. There's no prospect in the near future of sending spacecraft to visit exoplanets, so everything must be done with telescopes looking at light from afar. Consequently, we care most about exoplanets with liquid water right at the surface. These have the chance of a biosphere that pumps lots of gases into an atmosphere. In principle, such biogenic gases are detectable in the light from the planet and might indicate life.

In 1853, William Whewell noted that Earth's distance from the Sun allowed liquid water between what he called a 'central torrid zone' and an 'external frigid zone'. In the 1950s, the American astronomer Harlow Shapley (1885–1972) (who discovered the dimensions of our galaxy) also talked about a zone around stars where planets could have liquid water on their surface. If a planet is too far away from its host star it ices over, and if it's too close it's too hot for liquid water. Nowadays, the term *habitable zone* (HZ) refers to the region around a star in which an Earth-like planet could maintain liquid water on its surface at some instant in time. We specify a particular time because stars age and brighten or dim, so the HZ moves. In contrast to the HZ, the *continuously habitable zone* (CHZ) is the region around a star in which a planet could remain habitable for some specified period, usually the star's main sequence lifetime.

The width of the HZ around different types of main sequence stars, including the Sun, has been estimated. The inner edge is set by a planet's susceptibility to the runaway greenhouse effect, while the outer boundary is usually determined by the failure of greenhouse warming when carbon dioxide is cold enough to condense into clouds of dry ice or dense enough to scatter away sunlight. The latter was a potential problem for early Mars (Chapter 6). For the Sun, the HZ's inner edge is around 0.85–0.97 AU, while the periphery is 1.4–1.7 AU. The spread reflects the uncertain effects of water and carbon dioxide clouds at the inner and outer borders, respectively. For example, clouds might cool a close-in planet by reflecting sunlight, so the inner edge might be 0.85 instead of 0.97 AU. For the most optimistic outer boundary, Mars resides within the HZ. However, Mars's small size has led to a thin atmosphere that can't sustain liquid water. In other words, if Mars had been as big as the Earth, perhaps it might be habitable today. So planet size matters and the HZ is not the whole story for habitability but just a first guide.

Around other stars, the HZ is closer in or further out depending on whether a star is cooler or hotter than the Sun. For example, a K star, which is cooler (Fig. 1), would have a tight habitable zone from about the orbit of Mercury to that of the Earth. An even cooler M-type red dwarf would have the centre of its habitable zone at only 0.1 AU or so, well within the 0.4 AU orbit of Mercury.

In fact, the HZ of faint M dwarfs is so snug that planets will experience *tidal locking*, which occurs when gravity sets the rotation period. The same phenomenon makes one face of the Moon point toward the Earth. The Moon is tidally locked and spins on its axis once for every orbit around the Earth. The match between orbital and rotation periods, called *synchronous rotation*, is no coincidence. The Moon elongates slightly in the Earth–Moon direction, and if this orientation deviated, the Earth's gravity would twist the Moon's skewed bulges back into alignment.

A synchronously rotating planet in the HZ of an M dwarf would have one side in sunlight and the other in perpetual darkness. However, this doesn't make such planets uninhabitable. Although the night side would be cold, a moderately thick atmosphere can transport enough heat to warm the night side while cloud cover can cool the day side. Also, a large moon orbiting such a planet would have variable sunlight. So tidal locking in the habitable zones of M dwarfs is not a showstopper for life. That's encouraging because M dwarfs are the most common type of stars, making up about three-quarters of the total.

In recent years, assumptions defining the HZ have been questioned. Conventionally, the outer edge is determined by carbon dioxide condensation. But some large, rocky exoplanets might have thick hydrogen-rich atmospheres. We know from studying Titan and the giant planets that hydrogen and methane behave as greenhouse gases, and these condense at far colder temperatures than carbon dioxide. Also, on the inner edge, a water-poor planet with polar lakes rather than oceans might not succumb to a runaway greenhouse because it wouldn't have enough water to generate an atmosphere completely opaque to infrared radiation. Thus, the HZ might be wider than we think.

Is there a galactic habitable zone?

Some astronomers also argue that there's an optimal region in the galaxy for habitable planetary systems called the galactic habitable zone (GHZ). They note that the Sun is two-thirds of the way from the centre of the Milky Way, whereas planetary systems near the densely populated centre would be perturbed by supernovae or passing stars. At the other extreme, stars near the galaxy's edge are very poor in elements other than hydrogen or helium, and this might curtail planet formation. Astronomers have an odd convention (which sickens chemists!) of calling *all* elements other than hydrogen or helium 'metals'; the 'metal' content of a star is called its *metallicity*. Stars with giant exoplanets tend to be metal rich, and since giant exoplanets were the ones discovered first, it was initially thought that high metallicity was needed for planets to form in a nebula. More recently, no correlation with metallicity has been found in stars of Sun-like mass that have Super-Earths.

Another problem with the GHZ is that stars don't stay put. Recent research shows that stars wander across the galactic disc as a result of gravitational scattering by spiral arms. In any case, the exoplanets that will be scrutinized for signs of life in the foreseeable future will all be close, within a hundred light years of us. So whatever the validity of the GHZ, it's not a practical consideration for finding habitable planets any time soon.

Biosignatures, or how we find inhabited planets

Of course, the central question is whether we can find life. In 1990, NASA's *Voyager 1* spacecraft was heading out of the Solar System and looked back at the Earth from 6.1 billion km away. The famous picture it took is known as the 'Pale Blue Dot', in which Earth is so small that it's a single bluish pixel. If we knew nothing else, what could we deduce? If we said that the planet's colour was a mixture of blue oceans and white clouds that would just be guesswork. How can we do better?

Generally, to determine an exoplanet's properties and look for life, we need a planet's visible or infrared light spectrum. Different molecules and atoms absorb different frequencies in spectra. Thus, we can look for atmospheric gases such as oxygen or methane that can be produced by life. Such planetary features that indicate life are *biosignatures*. In fact, in the same year as *Voyager 1*'s 'Pale Blue Dot', the *Galileo* spacecraft, which was headed for Jupiter, obtained spectra of Earth. It was a sort of dry run for exoplanets and showed that you could detect Earth's atmospheric oxygen, methane, and abundant water. The simultaneous presence of oxygen and methane is evidence for life because these two gases should quickly react with each other unless a biosphere is producing great amounts of them, which prevents equilibrium. Thus, we say that Earth's atmosphere is in *chemical disequilibrium*. That's the kind of biosignature we would like from exoplanets.

Already, some information has been collected about exoplanet atmospheres by looking at differences in spectra when transiting exoplanets pass behind or in front of a star. The spectrum in the *primary transit* when a planet crosses the face of a star includes light passing through a ring of atmosphere around the planet. So subtracting the spectrum of just the star after the planet has passed by can isolate the spectrum of the planet's atmosphere. Alternatively, you take a spectrum when the planet is behind the star in its *secondary transit* and you subtract that from a spectrum of the planet plus star when the planet is beside the star. This produces the spectrum of the whole planet. In fact, this technique has been used to obtain infrared spectra of hot Jupiters with the Spitzer Space Telescope, which is a telescope orbiting the Sun.

With more sophisticated telescopes, we would like to determine an Earth-like exoplanet's surface temperature and whether it has liquid water. A spectrum might provide that, but you might only be able to see light from cloud tops or the upper atmosphere. In fact, Venus is veiled in this way. For exoplanets that are similarly obscured, we'll need to infer the amount of greenhouse gases from absorption lines in spectra and calculate a surface temperature. Together with an exoplanet's atmospheric composition, you could then infer whether liquid water is present. It might also be possible to look for an ocean through glint, which is the bright reflection spot on a smooth body of water at glancing angles. Finally, we should keep an open mind about *anti-biosignatures*. Microbial life will readily eat hydrogen or carbon monoxide gas, so an abundance of either of these might be considered an anti-biosignature for a planet in the habitable zone. Basically, anti-biosignatures say 'no one home'.

The search for extraterrestrial intelligence (SETI)

A different approach to finding life on exoplanets is to hypothesize that technological civilizations exist. If they do, we can look for their communications. This endeavour is the search for extraterrestrial intelligence (SETI) or more generally the search for *technosignatures*.

In 1959, Guiseppe Cocconi and Philip Morrison advocated looking for broadcasts of extraterrestrial civilizations using large radio telescopes. Others have since suggested watching for transmissions in visible light. There are practical reasons to look only for these signals. For example, X-rays are absorbed in the upper atmosphere, whereas radio and visible light are easy to generate and detectable across space. The main question for SETI is whether it's worth looking. Are there enough civilizations out there?

In 1961, Frank Drake (then at Cornell University) described a way to evaluate the potential number of transmitting civilizations in our galaxy. If we estimate the average number of civilizations that come 'on air' each year and their average lifetime, we can multiply those two quantities together to give the current number of transmitting civilizations. To give an analogy, if the number of new freshmen starting at a university is around 6,000 per year and they spend an average of four years on campus, the total undergraduate population at any one time is $6{,}000 \times 4 = 24{,}000$. Drake went further by devising six factors that determine the number of 'freshman' transmitting civilizations appearing per year. In full, we calculate N, the number of communicating civilizations, by multiplying the six numbers and the average lifetime, as follows:

N civilizations =

$$\underbrace{R_* \times f_{\text{planet}} \times n_{\text{habitable}} \times f_{\text{life}} \times f_{\text{intelligence}} \times f_{\text{civilizations}}}_{\text{number of communicating civilizations appearing each year}} \times \underbrace{l}_{\text{their average lifetime}}$$

This is the famous *Drake Equation*. The first number R_*, is the birth rate of stars suitable for hosting life. Astronomers observe about ten new stars per year of types G, K, and M. All the other factors are as follows:

f_{planet} is the fraction of such stars having planets;
$n_{habitable}$ is the average number of planets per planetary system that are habitable;
f_{life} is the fraction of those planets on which life originated and evolved;
$f_{intelligence}$ is the fraction of inhabited worlds that developed intelligent life;
$f_{civilizations}$ is the fraction of those worlds that developed civilizations capable of interstellar communication;
and l is the lifetime of those communicating civilizations.

Several factors in the equation are unknown. But, what the heck? Let's use the Drake Equation anyway. Current exoplanet searches suggest that at least two-thirds of all stars have planets, so let's take f_{planet} as 2/3. Data from the *Kepler* mission are still being analysed, but suggest at least that 1 in 100 of planets are habitable, so let's set $n_{habitable}$ = 1/100. We will simply have to guess all other parameters. Life developed rapidly on Earth, so let's presume that life originates on half of the habitable planets, i.e. f_{life} = 1/2. Let's suppose that the fraction of biospheres that develop intelligence is $f_{intelligence}$ = 1/8. Let's also guess that one in ten intelligent biospheres develop civilizations capable of interstellar transmissions, so $f_{civilizations}$ is 1/10. Finally, the lifetime, l, of communicating civilizations is sociological speculation, but let's say 10,000 years. So how many communicating civilizations in the Milky Way do we get? The answer is four, from $10 \times (2/3) \times (1/100) \times (1/2) \times (1/8) \times (1/10) \times 10{,}000$. Fine. But can we constrain the probabilities better?

Exoplanet studies will eventually nail down the second and third terms in the Drake Equation and might have something to say about the probability of life arising, f_{life}, if biosignatures are detected. Optimists think that a planet with liquid water and the right materials develops life easily. At present, we don't really know.

The next term concerns intelligence, so how probable is that? A relevant issue might be that only some biological solutions work to solve specific problems. In zoology, we often find the same trait shared by organisms in unrelated lineages, a result of what is called *convergent evolution*. Convergence occurs for specific functions and ecological niches. For seeing, eyes have evolved in at least forty different animal groups. A dog's life, of all things, is an example of a convergent niche. The Tasmanian wolf (a marsupial) evolved dog-like traits similar to the Mexican wolf (a placental mammal). Evolutionary convergence is so common that it means that organisms with legs, pairs of eyes, and certain ecologies might be inevitable because only these are physical solutions to the problems of walking, stereo vision, and occupying particular niches.

Whether technological intelligence is unique or convergent is contentious. On Earth, it was slow to appear. It took four billion years to build up sufficient O_2 to allow animal life. Then another few hundred million years passed before technological life. Dinosaurs reigned for about 170 million years and yet we find no sign of technology, neither fossil tools nor dinosaur microwave ovens. On the other hand, there are some lineages where brains appear to have had value and grown. Brain mass itself is a poor measure of intelligence because a big body often needs a big brain to run it. Instead, the Encephalization Quotient (EQ) is used, which is the ratio of the brain mass of an animal to the brain mass of an average animal of the same body mass. Thus, an average animal has an EQ of 1. Scores higher than 1 are brainy and scores below 1 more brainless (Fig. 11). Humans and chimps have EQs of about 7.4 and 2.5, respectively, which means that they have larger brains than expected by these factors. A rabbit has an EQ of 0.4, which would please Elmer Fudd. As Aristotle noted, 'man has the largest brain in proportion to his size'.

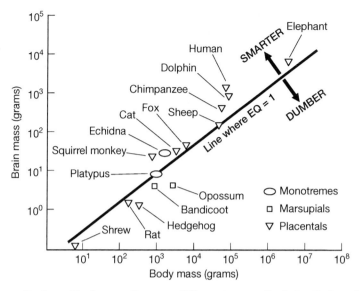

11. Brain and body mass for some different mammals. Animals that plot to the left of the diagonal line have more brain mass than typical

Palaeontologists have discovered an increase in EQ in certain lineages over time, most notably the genus *Homo*. The EQ of *Homo habilis*, a hominin that lived about 2 Ma, was only 4, for example. In toothed whales (dolphins, sperm whales, and orcas), EQ increased around 35 Ma, when they developed echolocation to find fish and friends. 'Friends' may be the key. Generally, intelligence is larger in social animals, such as 5.3 for a dolphin. Intelligence probably aids survival and getting a mate, which, in turn, promotes more offspring. However, there's much debate about which evolutionary pressures are behind intelligence.

A final question about SETI is called the *Fermi Paradox*, and was conceived by Enrico Fermi (1901–54), the Nobel Prize-winning physicist who built the world's first nuclear reactor in 1942. His idea arose from a lunchtime conversation. Stars in the Milky Way disc date from nine billion years ago, whereas the Earth is only 4.5 billion years old. Thus, if intelligent life is common, many technological civilizations should have arisen long before us.

Assuming that they developed space travel (or self-replicating robots to do the space travel for them), they should have spread throughout the galaxy by now. 'Where are they?' Fermi asked.

Fermi's Paradox has three solutions. One is that we're alone in the Milky Way because one factor in the Drake Equation is vanishingly small. A second is that Fermi's premise is incorrect. For all manner of reasons, civilizations might not go around colonizing the galaxy. Finally, there are civilizations but they're hiding their existence from us (or most of us). Scientists particularly dislike the third option because it's an ad hoc hypothesis that's untestable, but it's beloved of science fiction writers, tabloid newspapers, and (according to polls) one-third of American adults who believe that extraterrestrials have visited. The reasonable options are the first two. Obviously, all efforts to find life elsewhere bear upon the question of whether we're alone and the first solution. If SETI succeeds, we might also gain insight into other civilizations, if we could understand their signals. At the moment, the only known life is here, so, like Fermi, we can only wonder, 'Where are they?'

Chapter 8
Controversies and prospects

The Rare Earth Hypothesis

The big, unanswered questions of astrobiology generate controversy. One debate surfaced in 2000 when Peter Ward and Don Brownlee, my colleagues at the University of Washington in Seattle, published a bestselling book, *Rare Earth*. In essence, their *Rare Earth Hypothesis* is that the fortuitous circumstances that have allowed complex life on the Earth are so uncommon that Earth might harbour the only intelligent life in the Milky Way. Amongst their arguments were the good fortune of Earth being in the right place in the galaxy, having Jupiter in our Solar System to capture comets that might otherwise collide with the Earth, Earth's unusual recycling of volatiles by plate tectonics to keep the atmosphere going, the contingencies in obtaining an oxygen-rich atmosphere, and the luck of having a large Moon that stabilizes the Earth's axial tilt and so its climate.

Rare Earth was a polemic that railed against the *Copernican Principle*. The latter idea (named after Nicolas Copernicus, whose Sun-centred system knocked the Earth from its perceived place as the centre of the universe) holds that there's nothing special about our location. Astronomers note several factors in its favour. First, the Earth is surely one of many rocky planets in the universe. Furthermore, our Sun, a G-type star, is not special because around one in ten stars are G type. We also live in a humdrum location in the galaxy, along one of many spiral arms. Finally, our galaxy is unremarkable among many in the observable universe. As Stephen Hawking has put it, 'The human race is just a chemical scum on a moderate-sized planet, orbiting around a very average star in the outer suburb of one among a hundred billion galaxies.'

Advocates of the Copernican Principle also note how the great advances of science highlight the folly of assuming that we're special. We need think only of Galileo's *The Starry Messenger* (1610), which reported his observations of moons of Jupiter and the surface of the Moon, showing that these bodies were not heavenly perfections as supposed by theologian philosophers but explained by the same physics that we have on Earth. Similarly, Darwin's *Origin of Species* (1859) overturned the conceit that humans exist outside the rest of biology.

The problem with the Rare Earth Hypothesis is that it assumes too much knowledge about habitability, whereas, in reality, much is uncertain. Recently, for example, it's been discovered that the wandering of stars means that the galactic habitable zone is a less concrete concept than previously thought. The results of NASA's *Kepler* mission also show that planets around other stars are common, including plenty of Jupiter-size bodies. The argument that having a Jupiter-sized planet in just the right place lowers the rate at which comets or asteroids hit the Earth has also been challenged. Certainly, Jupiter mops up some impactors like a cosmic vacuum cleaner. In 1994, the comet Shoemaker–Levy 9 smashed into Jupiter, while in 2009 and 2010, the scars of further impacts were seen on Jupiter. It's well established that Jupiter helps to deflect and eject comets that come from a halo of such icy objects, the Oort Cloud, which surrounds the Solar System at about 50,000 AU distance. However, near-Earth asteroids and comets from within the plane of the Solar System represent more than 75 per cent of Earth's impact threat and Jupiter can actually destabilize those objects. So Jupiter is two-faced. Some calculations suggest that Jupiter is actually a net foe rather than friend.

Other arguments for Rare Earth are also ambiguous. In the Solar System, Earth uniquely has plate tectonics. However, for plate tectonics, a planet must be big enough to have adequate internal heat to drive the plate motion and it probably needs seawater to cool oceanic plates and lubricate their movement. In the Solar System, only the Earth qualifies. But that doesn't mean that Earth-like exoplanets in habitable zones might not also be suitable for similar tectonics.

While the Rare Earth Hypothesis is correct that planets without a large moon will suffer larger axial tilt variations than Earth, climatic variations at low latitudes might be benign. In fact, thicker atmospheres, more extensive oceans, and lower rotation rates of an exoplanet can smooth the climatic differences between pole and tropics caused by a varying tilt. Finally, the question of oxygen not accumulating on other Earth-like planets might go in the other direction to that assumed in the Rare Earth Hypothesis. Some planets might be more favourable for oxygen-rich atmospheres than Earth because their volcanoes pump out a smaller proportion of gases that react with oxygen. Oxygen might build up more easily.

What does seem to be correct about the Rare Earth Hypothesis is that microbial-like life should be much more common than intelligent life. Microbes have a remarkable range of metabolisms and can live in a far wider variety of environments than complex organisms. But any definitive statements about the prevalence of complex life—one way or the other—simply lack data to support them and we should be sceptical. As Carl Sagan famously remarked, 'It pays to keep an open mind, but not so open that your brains fall out.'

Prospects for astrobiology and finding life elsewhere

The excitement of astrobiology is that it tries to answer questions such as the origin of life and whether we're alone in the universe. With advances in technology, it's increasingly likely that major discoveries will be made in the coming decades.

In the area of understanding early life, it's likely that realistic self-replicating genomes will be made in the lab. This would provide great insights into the origin of life. Several groups are studying the RNA World or its variants. There are also projects to drill deeply into old sedimentary rocks in South Africa and Australia, which will surely make new discoveries about the earliest life and environment.

In the Solar System, Enceladus ought to be one of the highest priorities for the world's space agencies. Enceladus has a source of energy (tidal heating), organic material, and liquid water. That's a textbook-like list of those properties needed for life. Moreover, nature has provided astrobiologists with the ultimate free lunch: jets that spurt Enceladus's organic material into space. Technology certainly exists to build a spacecraft to swing by Enceladus and sample the organics in the jets. Better still, the material could be returned to Earth for analysis.

In fact, spacecraft to collect extraterrestrial samples and return them to Earth, which are *sample return* missions, are the future in understanding the history of Mars and Venus and whether either of these planets was once inhabited. A sample return mission for Venus is in the distant future, but one for Mars is a strategic goal of NASA's and ESA's current programmes.

Around Jupiter, Europa is probably the best prospect for life. The first step would be a Europa orbiter to study the moon in detail and determine the thickness of the ice above a subsurface ocean or lake lenses. The next steps might involve landers, and possibly robots to melt through the ice using radioactive heat generators. Eventually one imagines submarines diving though a Europan ocean.

Apart from Enceladus, Titan is a target of astrobiology amongst Saturn's moons. A huge scientific leap could be made if a lake lander—a sort of interplanetary boat—could float on the lakes in the polar regions of Titan and find out which substances make up the organic liquids. Furthermore, a Titan orbiter could do the kind of reconnaissance that would determine the depth of Titan's subsurface ocean and study Titan's surface.

One certainty for the future is that exoplanet discoveries will continue to spur an interest in astrobiology. I anticipate the discovery of many Earth-like planets inside the habitable zone of other stars, dead planets with almost pure carbon dioxide atmospheres, water worlds covered entirely in glinting oceans, and young Venus-like planets sweating off their oceans into space from runaway greenhouse effects.

When astrobiology came to the fore as a discipline in the 1990s, some questioned its future and wondered if it might be a fad that fades, perhaps because of disappointment in not quickly finding extraterrestrial life or a failure to answer questions about life's origin. However, the discovery of Earth-sized exoplanets in habitable zones will ensure that the possibility of life elsewhere becomes more relevant than ever. Astrobiology is here to stay.

Further reading

Chapter 1: What is astrobiology?

Much lengthier introductions to astrobiology are found in the following textbooks:

J. O. Bennett, G. S. Shostak. *Life in the Universe*. (San Francisco: Pearson Addison-Wesley, 2012).

D. A. Rothery et al. *An Introduction to Astrobiology*. (Cambridge: Cambridge University Press, 2011).

K. W. Plaxco, M. Gross. *Astrobiology: A Brief Introduction*. (Baltimore: Johns Hopkins University Press, 2011).

J. I. Lunine. *Astrobiology: A Multidisciplinary Approach*. (San Franciso: Pearson Addison Wesley, 2005).

W. T. Sullivan, J. A. Baross (eds). *Planets and Life: The Emerging Science of Astrobiology*. (Cambridge: Cambridge University Press, 2007).

The development of astrobiology from the 1950s onwards is described by: S. J. Dick, J. E. Strick. *The Living Universe: NASA and the Development of Astrobiology*. (New Brunswick, NJ: Rutgers University Press, 2004).

An old classic on the nature of life is: E. Schrödinger. *What Is Life?* (1944; Cambridge: Cambridge University Press, 2012).

Chapter 2: From stardust to planets, the abodes for life

A readable discussion of modern Big Bang theory is given by:

C. Lineweaver, T. Davis. 2005. Misconceptions about the Big Bang. *Scientific American* 292: 36–45.

A popular account of Arthur Holmes's quest to find the age of the Earth is: C. Lewis. *The Dating Game: One Man's Search for the Age of the Earth*. (Cambridge: Cambridge University Press, 2012).

Chapter 3: Origins of life and environment

The science of the origin of life is described by: R. M. Hazen. *Genesis: The Scientific Quest for Life's Origin*. (Washington, DC: Joseph Henry Press, 2005).
A lucid description of the early evolution of life on Earth is: A. H. Knoll. *Life on a Young Planet: The First Three Billion Years of Evolution on Earth*. (Princeton: Princeton University Press, 2003).

Chapter 4: From slime to the sublime

The Earth's formation, evolution, and habitability are covered in: C. H. Langmuir, W. S. Broecker. *How to Build a Habitable Planet: The Story of Earth from the Big Bang to Humankind*. (Princeton: Princeton University Press, 2012).

Chapter 5: Life: a genome's way of making more and fitter genomes

A widely used introductory textbook on modern microbiology is: M. T. Madigan et al. *Brock Biology of Microorganisms*. (San Francisco: Benjamin Cummings, 2012).
The effects of life on the Earth's chemistry on a global level are described in the following textbook: W. H. Schlesinger, E. S. Bernhardt. *Biogeochemistry, Third Edition: An Analysis of Global Change*. (San Diego: Academic Press, 2013).

Chapter 6: Life in the Solar System

The planets of the Solar System and their habitability are described in the following textbook: J. J. Lissauer, I. de Pater. *Fundamental Planetary Science: Physics, Chemistry and Habitability*. (Cambridge: Cambridge University Press, 2013).

Chapter 7: Far-off worlds, distant suns

A readable book that discusses the search for habitable exoplanets is: J. F. Kasting. *How to Find a Habitable Planet*. (Princeton: Princeton University Press, 2010).

Chapter 8: Controversies and prospects

The controversial but engrossing book that argues for the scarcity of complex life is: P. D. Ward, D. Brownlee. *Rare Earth: Why Complex Life is Uncommon in the Universe*. (New York: Copernicus, 2000).

"牛津通识读本"已出书目

古典哲学的趣味	福柯	地球
人生的意义	缤纷的语言学	记忆
文学理论入门	达达和超现实主义	法律
大众经济学	佛学概论	中国文学
历史之源	维特根斯坦与哲学	托克维尔
设计，无处不在	科学哲学	休谟
生活中的心理学	印度哲学祛魅	分子
政治的历史与边界	克尔凯郭尔	法国大革命
哲学的思与惑	科学革命	民族主义
资本主义	广告	科幻作品
美国总统制	数学	罗素
海德格尔	叔本华	美国政党与选举
我们时代的伦理学	笛卡尔	美国最高法院
卡夫卡是谁	基督教神学	纪录片
考古学的过去与未来	犹太人与犹太教	大萧条与罗斯福新政
天文学简史	现代日本	领导力
社会学的意识	罗兰·巴特	无神论
康德	马基雅维里	罗马共和国
尼采	全球经济史	美国国会
亚里士多德的世界	进化	民主
西方艺术新论	性存在	英格兰文学
全球化面面观	量子理论	现代主义
简明逻辑学	牛顿新传	网络
法哲学：价值与事实	国际移民	自闭症
政治哲学与幸福根基	哈贝马斯	德里达
选择理论	医学伦理	浪漫主义
后殖民主义与世界格局	黑格尔	批判理论

德国文学	儿童心理学	电影
戏剧	时装	俄罗斯文学
腐败	现代拉丁美洲文学	古典文学
医事法	卢梭	大数据
癌症	隐私	洛克
植物	电影音乐	幸福
法语文学	抑郁症	免疫系统
微观经济学	传染病	银行学
湖泊	希腊化时代	景观设计学
拜占庭	知识	神圣罗马帝国
司法心理学	环境伦理学	大流行病
发展	美国革命	亚历山大大帝
农业	元素周期表	气候
特洛伊战争	人口学	第二次世界大战
巴比伦尼亚	社会心理学	中世纪
河流	动物	工业革命
战争与技术	项目管理	传记
品牌学	美学	公共管理
数学简史	管理学	社会语言学
物理学	卫星	物质
行为经济学	国际法	学习
计算机科学	计算机	化学
儒家思想	亚当·斯密	天体物理学
未来	天体生物学	